BLOCKCHAIN BASICS BIBLE

Non-Technical Beginner's Introduction to Cryptocurrency. The Future of Bitcoin & Ethereum Crypto Technology, Non-Fungible Token (NFT), Smart Contracts, Consensus Protocols, Mining, & Blockchain Gaming

By

Satoshy Nakamoto

DISCLAIMER

© Copyright 2021

Table of contents

INTRODUCTION

WHAT IS BLOCKCHAIN?

Individuals can collectively uphold a database without having to depend upon a central authority via Blockchain. Aimed for our unified future, Blockchain is prototypical for sharing and reconciling information.

Information has been traced through a ledger for a long time which is a simple sequential list of data. Ledgers have been used to track heaps of wheat, deaths from epidemics or from curses, basically any thing man can think of.

Ways of tracking and syncing data have extended with the development of the scope of human activity. With the development of computers and other digital resources in the previous century, our information also became electronic and the ledgers were now recorded on computers. The number of prospects and databases increased, and with the invention of new methods of improved, interconnected, electronic ledgers, it was suddenly easy to search, sort, share, and transport information.

Today, almost every digital service is reinforced by databases. But the practicality of database is being restricted, and its universality is being questioned due to it being originated in the pre-internet and pre-global world. Databases are sustained by a dominant authority that has complete power over the information the database contains. For instance, Facebook could change your Instagram to be just pictures of pretty flowers if it wanted to, and you'd have no choice but to accept it. Similarly, if a bank wanted to they could easily take $100 from your checking account. You'd clearly notice this and complain. But what if they only took a dollar per day?

We need something more irrepressible and cooperative, something that is wide-ranging but not subjected to a single dogmatic ideology, private intentions, or corporate incentive to manage our data since the world and our interactions have become more linked, and more digital.

Blockchain, invented in 2009, is exactly what we needed.

Blockchain can keep data coordinated across numerous, liberated stakeholders. Blockchain allows a collection of unrelated entities, having an enticement to modify their shared data, to agree on and maintain a unified dataset. In contrast to the the blockchain, a traditional database is better equipped for tracking the records of a single entity.

So far, Blockchain has been the most convenient for tracking money because the reason to edit data in your own favor is very strong in financial systems. The basic idea of how blockchain works are given below.

The Basic Idea Behind Blockchain

First, by running the same blockchain software, computers wanting to share data join together on a network. The data is assembled together into "blocks" for authentication as it is coming into the network; for instance through people spending and receiving money. The computes vote on the current block of data systematically, typically every few minutes or even every few seconds, deciding whether it all looks good or not. The computer votes on the overruled current block again when the next block is submitted. When the computer agrees about the validity of the data it holds, the current block is accepted, and then added to the system's complete past history of authenticated data blocks. Hence, the data is "chained". A long chain of information is formed as a result.

Every computer on the network stores the chain, cryptographic functions are used to carry out the appending, hence making it easy to tell if any past transaction has been changed even in the slightest. So with the addition of each new block of data, the whole network confirms the authenticity of all preceding data.

Let's say if you have more money than you're supposed to and wanted to cheat and, say, modify any past transaction, you'd have to change the histories across all the computers on the system working independently. Or have sufficient new computers join the network that could vote your cheats into "fact." Such an attack is called a 51% attack. Such an attack would be ridiculously costly if you're a part of a large system like Bitcoin because you'd have to run too many computers. Such attacks are unmanageable, by design, on Stellar. You can't even submit bad data instantaneously because ultimately the network will approve the right data and your self-seeking objective will be voted down.

As expected, a reasonable amount of technical detail goes into all this. We highly recommend reading the primary paper describing the first blockchain system, because it is rather ingenious.

Important Blockchain Products

Blockchain is not looked at as a competition at Stellar. There's a lot of good software out there! If you are looking to gain better information and knowledge of blockchain technology, you can resort to the following resources.

Bitcoin

Almost a decade ago the yet unidentified "Satoshi Nakamoto" created Bitcoin; a foundation and inspiration for essentially all blockchain systems. The up-front offshoots of BTC include popular platforms like Litecoin and Dogecoin. Even though much of its fundamental principles are the same, Stellar doesn't use the same technology.

This is a good rudimentary technical/conceptual Bitcoin primer.

This is another good indication which draws in some of what inspired the creation of the technology.

One of the plus points of a decentralized currency is presented by this Op-Ed by Meridian speaker Carlos Hernandez. Bitcoin has a tangible use case as an economic sanctuary in many parts of the world, despite it being a curiosity for Americans. Stellar anticipates building something that is as dominant but more accessible and like-minded with the present monetary system.

Ethereum

The second-largest blockchain platform is Ethereum. Even though it is somewhat similar to Bitcoin, it was thought of as something more multi-purposed than being just a currency like Bitcoin. Designed to become another Internet-style network or "world computer", Ethereum has its own programming language (Solidity) that ideally enables you to create any type of program inside the Ethereum network. These programs are known as decentralized applications or "dapps", and would even be ideally sturdy to government control and inaccuracy.

The Ethereum blockchain network has smart contracts as one of its main elements. Smart contracts are a type of contract that works automatically. These tokens are issued via Solidity. At this point, the spirit of Ethereum surpasses its capabilities. This is a reason that the Ethereum network is pretty slow. The intricate nature of Solidity is making it easy to exploit and it adds to the problem as well. It has its flaws and defects but is still considered to be one of the best examples of programming models.

A basic and ideal up to a certain level is Ethereum primer.

The DAO – the "Digital Autonomous Organization" was one of the firsts and ideal Ethereum projects that superbly went sideways.

Stellar

The creation of Stellar happened in the year 2014. It was after the bitcoin but before the creation of Ethereum. Stellar was created in order to facilitate remittances and payments. Stellar works on an environment-friendly syncing mechanism. No platform worked on such mechanism. Stellar is like a cashlike which causes short suspensions in between the transactions. Similar to that of Ethereum you can issue other assets and then trade them with great ease on Stellar.

Stellar can be connected to real-world endpoints. This is something that allows people to convert their digital money into the form that they can spend. This is something that you cannot do on every blockchain system. The growth of this network is something that the Stellar Development Foundation is working tirelessly on.

The Stellar community site which goes by the name Lumenauts.com has a lot of great Stellar interpreters. They have come up with a course that provides detailed information regarding the entire platform.

A few other projects worth checking out

Z-Cash

Z-Cash's tech is derived from Bitcoin. It is altered so it is able to guard and secure the users' privacy. Unlike the Ethereum Blockchain network that stores all the information on a digital ledger that can be traced by anyone, Z-Cash keeps some of the information of the transactions anonymous.

Basic Attention Token

This is an Ethereum-based project which is considered as an example. They aim to re-evaluate the working of internet advertising by building a resourceful and proficient "attention economy" using their token and their Brave browser.

0x

0x is a decentralized exchange protocol that lets people trade their coins peer-to-peer and not have any exchange on its own in the middle. It is something that happens in the CoinBase. It is a project by Ethereum. An example of this can be RadarRelay.

CHAPTER NO 1

THE HISTORY OF BLOCKCHAIN TECHNOLOGY

Blockchain technology is a decentralized, distributed ledger that records information on the transactions. The in-built design does not allow any data to be altered on the blockchain. It makes it a genuine disruptor for industries like payments, cybersecurity, and healthcare. This guide will help you understand how can it be used and what is its history.

Blockchain is also known as Distributed Ledger Technology (DLT). The digital ledger technology allows the history of any digital asset unable to change. It is kept transparent by using decentralization and cryptographic hashing.

Google Docs is a great way to understand Blockchain technology. Whenever a document is created on Google Docs, it can be shared with people. You see that the document that was created is distributed and not copied or transferred. This helps create a decentralized distribution chain that provides everyone access to the document simultaneously. No one has to wait for somebody else to make the changes that they want to make. All the changes that are made to the document are recorded instantaneously. This makes the whole process transparent.

Yes, blockchain is pretty complicated in comparison to a Google Doc, but the correspondence is appropriate because it demonstrates three critical ideas of the technology.

Risk reduction is one of the great factors that makes Blockchain a promising and groundbreaking technology. It is a great way to eliminate any fraudulent situations and ensures transparency in an accessible way for a myriad of uses.

How Does Blockchain Work?

Blockchain is used so that people can share important data in a secure, tamperproof way. — MIT Technology Review

The three important concepts that blockchain consists of are: blocks, nodes, and miners.

Blocks

Each of its chains is a composition of multiple blocks. Each block further has three basic elements.

The data in the block.

Nonce is a 32-bit whole number. It is randomly generated whenever a block is created, which in turn generates a block header hash.

Hash is a 256-bit number linked to the nonce. It needs to start with an enormous number of zeroes (i.e. be extremely small).

Nonce is a cryptographic hash that is generated when the first block of the chain is created. The data in the block is deliberated, signed, and forever tied to the nonce and hash till it is mined.

Miners

New blocks of data are formed by the miners on the chain. This process is termed mining.

Unique nonce and hash are a part of every block that is a part of the blockchain. Not just that but the nonce and hash that are used in the previous blocks in the chain are mentioned as well. This shows that mining a block is not something easy, particularly on large chains.

Miners use special software that is used to solve the extremely complex math problem of finding a nonce that generates an accepted hash. The size of nonce is 32 bits and that of hash is 256. This means that there are approximately four billion nonce-hash combinations that must be mined before the right one is initiated. After this, the miners are said to have established the "golden nonce." Once done that block that they just created is added to the chain.

You need to re-mine the block in which the change is made as well as all those blocks that come after that. This is done when changes are made in the blocks early. This makes blockchain technology very hard to influence. Think of it as "safety in math." That is because finding golden nonce entails a lot of time and power of computation.

All the nodes that are in the chain recognize and accept the changes that are made once a block is mined. This is a great way for the miners to earn some extra money.

Nodes

Decentralization is considered to be one of the most important concepts that you need to understand regarding blockchain technology. The chain is not owned by a specific organization or a computer. It is a distributed ledger by the nodes linked to the chain. Any electronic device that retains the copies of the blockchain and keeps it running at all costs is known as a node.

Every node that is part of the network owns its own copy of the blockchain. It is significant for the network to algorithmically accept any block that is mined for the chain to be updated, trusted, and verified. Blockchains are transparent. Everything that happens in the ledger can be checked and viewed with ease. In order to display the number of transactions, every member is given an ID that is alphanumeric and unique in nature.

The combination of the checks-and-balances system and public information helps the blockchain maintain the integrity and creates trust among its users. Blockchains are a great way to gain trust through technology.

Cryptocurrencies are a popular application of blockchain technology. Cryptocurrencies are also known as digital currencies (and sometimes tokens) such as Bitcoin, Ethereum, or Litecoin. These digital currencies are used to buy goods and services. It is said to be cash, but in a digital form. You can use crypto to purchase everything. Crypto is based on a blockchain system that is both a public ledger and an improved cryptographic security system. This is something that is not their cash. The transactions are recorded and kept secure.

There are 6,700 cryptocurrencies in the world to date with a total market capitalization of around $1.6 trillion. Bitcoin has a major value. These tokens have become extremely prevalent for a few years. The worth of a bitcoin is equivalent to $60,000. But why have people started to notice cryptocurrencies all of a sudden? Here are a few reasons for that:

Theft and fraud get difficult because each cryptocurrency has its own indisputable and recognizable number that is committed to one owner. This is a plus point of Blockchain security.

Crypto has decreased the want for individualized currencies and central banks. Crypto can be directed anywhere due to blockchain technology. That too, without any requirement for exchanging currency and with no central banks interfering in the matter.

Cryptocurrencies have the ability to make people earn more. Investors have been fueling the Crypto's price, especially Bitcoin. The people who adopted this technology in their initial days have now become billionaires. Whether this is actually a positive has yet to be understood as some retractors believe that investors do not have the long-term profits of crypto.

The majority of the large corporations have started to accept the idea of a blockchain-based digital currency for payments. Tesla has already made an investment of $1.5 Billion into Bitcoin back in February 2021. They have accepted it as payment for their cars.

There are multiple points of view regarding the digital currencies that are based on the blockchain. The crypto market is not much regulated. Many governments jumped onto the crypto bandwagon, but few have a firm set of codified laws regarding the crypto market. Due to its investors, the crypto currencies are highly volatile. In 2016, Bitcoin was estimated to be around $450 for each token. It then soared to $16,000 per token in the year 2018. Then it plunged to around $3,100. An increase was seen and it got to $60,000. This absence of stability has triggered some people to get very rich, while a majority has lost thousands.

If we state the prerogative that cryptocurrencies are the future will be something ahead of time. For now, the swift rise in blockchains is preliminary to take essence in reality than buildup. Blockchain still looks like a promising field as compared to Bitcoin.

Applications of Blockchain in the Music Industry:

The growth of technology and the internet has paved the way for the music industry. It has gone huge over the past few years. It has been noted that that there has been a significant increase in music streaming websites and platforms. Artists, labels, publishers, songwriters to the streaming service providers all are being influenced negatively in the music industry due to this. Music royalties are determined by an intricate process but the internet has made it even more worse by giving rise to the demand for transparency in the royalty payments by artists and songwriters. This is where blockchain technology comes into play. It maintains an all-inclusive, precise distributed database of music rights ownership information in a public ledger. Furthermore, the rights to ownership information, royalty splitting as determined by "smart contracts" could be made a part of the database. The relationships between the different stakeholders can be programmed by the use of smart contracts.

Dropbox and Google Drive are Decentralized Storage Cloud file storage solutions that are seeing growth in popularity. They are used to store documents, photos, videos, and music files. Notwithstanding their popularity, cloud file storage solutions characteristically face challenges in areas such as security, privacy, and data control. One of the biggest issues that one comes across in these scenarios is that you have to trust a third party with one's confidential files. Storj provides a blockchain-based peer-to-peer distributed cloud storage platform that permits users to transfer and share data without trusting a third-party data provider. This lets the people share the internet bandwidth that is not used as well as extra disk space in their devices. This is something that would help those people who are considering storing large files in return for micropayments based on the bitcoin. Lack of central control removes data failures that are more conventional. This also considerably increases security, privacy, and data control. Storj platform can occasionally cryptographically check the integrity and availability of a file, and offer direct rewards to those maintaining the file. Micropayments that are based on the incentive and payment are served in a separate blockchain. That is used as a datastore for file metadata.

Beyond Bitcoin: Ethereum Blockchain

The extremely transparent ledger system for Bitcoin that it works on, and blockchain have been linked with cryptocurrency. But then again the transparency and security of the technology have been flourishing in a number of zones. The majority of this can be traced back to the development of the

Russian-Canadian developer Vitalik Buterin in the late months of 2013 who published a white paper. The white paper projected a platform that combines the working of traditional blockchain with a major alteration: the execution of computer code. This gave birth to the Ethereum project.

Ethereum blockchain allows the developers to create sophisticated programs that have the ability to communicate with each other on a blockchain.

Tokens

Tokens are created by the Ethereum programmers to signify the digital assets. Not just that, but coming up with trajectories to find who owns the assets as well, and execute its functionality according to programming instructions.

Tokens can be anything from music files, contracts, concert tickets, or even a patients medical records. Nowadays the popularity of the Non-Fungible Tokens (NFTs) has increased multiple folds as well. NFTs are exclusive tokens based on blockchain technology that are used to collect digital media or art. NFTs have the capability to prove and authenticate the past history and sole ownership of the piece of digital media. NFTs have greatly helped the artist in selling their digital artworks and also get proper credit and a fair share of profits.

The applications of blockchain have expanded the likelihood of the ledger technology to saturate other areas like media, government, and identity security. Several companies are working on developing products and ecosystems that run completely on the rapidly increasing technology.

Blockchain is inspiring the existing state of affairs of novelty. They are allowing the companies to experiment with revolutionary technology like peer-to-peer energy distribution or decentralized forms for news media. This is pretty much how Blockchain is defined. With the evolution of technology, the evolution of the ledger system would happen too.

History of Blockchain

Despite being a new technology, blockchain already claims a rich and interesting history. A lot of important events occurred in the past and in order to distinguish events that led to the development of blockchain are given below.

2008

In 2008, a book called "Bitcoin: A Peer to Peer Electronic Cash System" was published by Satoshi Nakamoto - the identity of whom is not known.

2009

In 2009, the first effective Bitcoin (BTC) transaction took place between computer scientist Hal Finney and the mysterious Satoshi Nakamoto.

2010

The first-ever purchase using a bitcoin – two Papa John's pizzas, was made by Florida-based programmer Laszlo Hanycez. A total of 10,000 BTC's worth $60 at that time, were transferred by Hanycez. Today it holds a value of $80 million.

Within the same year, the market cap of Bitcoin publically surpassed $1 million.

2011

The cryptocurrency was given equivalence to the US dollar with 1 BTC being equal to $1 USD.

Bitcoins started being accepted as donations by organizations such as Electronic Frontier Foundation, Wikileaks.

2012

Blockchain was introduced into pop culture when renowned TV shows like The Good Wife mentioned cryptocurrency and Blockchain.

Early Bitcoin developer Vitalik Buterin launched Bitcoin Magazine.

2013

BTC's market cap exceeded $1 billion.

The value of Bitcoin equivalent to dollar reached the value of $100 BTC for the first time.

"Ethereum Project", the paper suggesting that blockchain had other possible uses besides Bitcoin, like smart contracts,was published by Buterin.

2014

Bitcoin has started being accepted as payment by gaming company Zynga, The D Las Vegas Hotel, and Overstock.com.

Buterin's Ethereum Project was crowdfunded through an Initial Coin Offering (ICO) which raised over $18 million in BTC and increased opportunities for blockchain.

A group called R3 was formed to discover new ways of implementing blockchain in technology. This group comprised over 200 blockchain firms.

Bitcoin integration was announced by PayPal.

2015

The number of merchants accepting BTC exceeds more than 100,000.

In order to test the technology for trading shares in private companies, NASDAQ and San-Francisco blockchain company Chain teamed up.

2016

A blockchain strategy for cloud-based business solutions was announced by the tech giant IBM.

The validity of blockchain and cryptocurrencies was recognized by the Japanese government.

2017

For the first time, Bitcoin reached $1,000 BTC.

The cryptocurrency market cap crossed $150 billion.

Jamie Dimon who is the Chief Executive Officer of JP Morgan specified his belief in blockchain as a future technology which gave the ledger system a vote of confidence from Wall Street.

Bitcoin reached its unsurpassed high at $19,783.21 BTC.

It was announced by Dubai that by 2020 its government will be powered by Blockchain.

2018

Facebook pledged to start a blockchain group and also suggested the possibility of creating its own cryptocurrency.

A blockchain-based banking platform was developed by IBM and great banks like Citi and Barclays signed on.

2019

Blockchain was publicly embraced by the Chinese president Ji Xinping as China's central bank announced it is working on its own cryptocurrency.

Twitter & Square CEO Jack Dorsey announced that Square will be hiring blockchain engineers to work on the company's future crypto plans.

The creation of Bakkt, a digital wallet company that comprises crypto trading was announced by the New York Stock Exchange (NYSE.)

2020

By the end of 2020, Bitcoin almost reached $30,000.

It was announced by PayPal that they would allow users to buy, sell, and hold cryptocurrencies via Paypal.

"Sand Dollar", the first central bank digital currency, was launched by the Bahamas, hence becoming the world's first country to do so.

Due to securely storing medical research data and patient information, Blockchain became a key player in the battle against the Corona Virus.

CHAPTER NO 2

HOW DO CRYPTOCURRENCIES RISE UP?

Just like a band, cryptocurrency is a digital currency that can be exchanged between two persons directly without the involvement of a third person. It shows the financial amount without disclosing the identities of the people carrying out the transaction, hence enabling consumers to digitally connect directly through a transparent process. In order to confirm a cryptocurrency exchange and prevent replication of the very transaction, the network consists of a chain of computers. This form of a transaction has the likelihood to reduce scams because of its transparency.

A cryptocurrency exchange is fairly analogous to PayPal, with the exception that the currency being exchanged is not traditional money. To guarantee the security of transactions, cryptocurrency uses digital safeguards. A process known as mining a digital public ledger, called a blockchain, must confirm each transaction.

How do Cryptocurrency Exchanges Work?

In order to clear up any confusion let's discuss the following words:

Transaction: The transfer of currency between two digital wallets is known as a transaction. Before the give-and-take can be settled, a transaction is submitted to a public ledger to await authorization. During a transaction in demand to show ownership, an encrypted electronic signature based on a mathematical formula is obligatory. People called miners carry out the process of confirmation.

Public Ledger: A public ledger called a blockchain store the transaction after a miner confirms it. The confirmation of the ownership and guarantee of the validity of recordkeeping is carried out by the public ledger.

Mining: Before being added to the public ledger, the transactions are confirmed by a process called mining. In order to prevent misuse of cryptocurrency mining, a miner must know how to solve "proof of work"; a computational puzzle. Before the transaction block is added to the blockchain, anyone can authorize the transaction, hence mining is open source. For their work in cryptocurrency miners receive a fee.

In short, a cryptocurrency exchange using blockchain practically works in the following ways:

1. "A" desires to send cryptocurrency to "B."

2. The exemplification of transactions online is known as a block.

3. Everyone on the network receives the block.

4. A miner within the network will approve the validity of the block.

5. The newly created block becomes a part of the public ledger or blockchain.

6. Then the currency moves from "A" to "B."

In traditional banking systems, intermediaries must be trusted by the sender and receiver to assist centralized transactions. Not only does this kind of demand cost great fees, but it takes hold of the private data of individuals while doing so. Conversely, cryptocurrency exchange safeguards individual identities whilst providing a decentralized, transparent mechanism for transferring value at a lower cost.

This guide will enable us to better understand the time when the cryptocurrency market was at its best. It will also allow us to gain some knowledge regarding the ICOs as well as the different types of cryptocurrencies.

The Rise of the Cryptocurrency Market

Largely anonymous to the world's wide-ranging population, ten years ago cryptocurrencies were an academic concept. This was all revolutionized in 2009 with the creation of Bitcoin. People may not be aware of how the system works, but today mostly know of cryptocurrencies.

The grip of the cryptocurrency market has become strong. Various aspects of government, business, and other personal financial activities are covered using cryptocurrencies:

—For the assessment of how the transaction mechanism, specifically blockchain technology, can be adapted to exchange value, government and large corporations are now looking into the cryptocurrency market.

—In order to evaluate the likelihood of incorporating this technology into their businesses many companies have begun blockchain projects.

— Just like the internet connects people from all over the world and facilitates data exchange, blockchain technology is considered to be the second type of internet by experts: the internet of value.

People today rely more on technology to provide secure services in a more well-organized and profitable manner. Financial service providers in particular are looking at cryptocurrency. Before considering the probable growth of the cryptocurrency market, let's talk about the start of it all:

The advent of Bitcoin as a standard in the market:

— Using centralized control many people tried to create digital currencies in the 1990s, but they all failed due to numerous reasons.

— A peer-to-peer cash system, called Bitcoin was developed in late 2008 by Satoshi Nakamoto. For the very first time, someone was able to build a protected, decentralized digital cash system.

—Double spending was prevented by Satoshi Nakamoto's system, conventionally something that only a centralized server could achieve. The foundation of cryptocurrency was based on Nakamoto's innovation.

— A decentralized network functions on a system of checks and balances, where every unit within the network checks to see if there is any attempt to spend the same currency twice. With the advent of bitcoin, it became easy to reach consensus without a central authority when no one deemed it possible.

For issuing currency, processing exchanges, and verifying transactions, Bitcoin uses peer-to-peer network and blockchain technology since it is a decentralized currency. Due to this, it is free of government meddling or manipulation; contrasting to a fiat currency, which is managed by a country's central bank.

The current rate of creation of Bitcoins via the mining process is 25 Bitcoins every 10 minutes. It is expected that in 2140 the number of Bitcoins in circulation will be stopped at 21 million. The disadvantage to cryptocurrency exchange is that the currency's value depends upon the demand from investors, and with the drop of the market, the value of Bitcoin drops as well.

Moreover, unlike money can in traditional banking systems, debts are not represented in cryptocurrencies. It is hard currency; as treasured as holding gold coins. As stated, the volume of Bitcoin is set at 21 million, so there's a limit set on the supply of cryptocurrency tokens.

The Transactional Characteristics of Cryptocurrencies

A number of characteristics of cryptocurrency transactions contrast from customary banking.

It is anonymous. It is not necessary for both the parties that are a part of the transaction to know and recognize each other but still the transaction is transparent. This has grabbed the attention of U.S. Federal agencies such as the FBI and the Securities and Exchange Commission (SEC), who are apprehensive about the prospect of money laundering.

It is secure. The person who owns the private key that allows them to access the funds can exchange cryptocurrencies, and the funds are safely locked in a system.

It is fast and worldwide. Geographical location is not a hindrance to allowing a transaction since the network operates worldwide. It takes only a few minutes for transactions to be mined and confirmed, making them faster than traditional banking systems.

It is irreversible. A transaction cannot be reversed after being confirmed and added to the blockchain. There is no alternative in the event that cryptocurrency is sent in error.

It does away with red tape. For using the cryptocurrency exchange system one does not need permission. It is free to download and use.

The Future of Cryptocurrencies

If the success of Bitcoin is any hint, the cryptocurrency market has an optimistic future even though it is not probable to foresee the future prospects of all of the cryptocurrencies.

— Price of bitcoin gradually increased from over $280 in July of 2015 to a mark of $1000 in January of 2017.

— Since then, remarkable growth has been recorded in cryptocurrency. The worth of Bitcoin got $17,000 by early December 2017. The price of "Ether", another cryptocurrency, has also continued to increase recently.

Initial Coin Offerings (ICOs) have played an important part in creating interest in the cryptocurrency market. Coins or tokens similar to a company's shares are used by ICO which are then vented to those who invested in an (IPO) transaction. IPO refers to Initial Public Offering. Equated to crowdfunding, ICO uses cryptocurrencies as a source of capital for startup companies. A cryptocurrency crash is anticipated by market experts at some point. It is foreseeable that a regulator like the SEC will want to intervene by providing guidance and putting into effect actions where necessary, when market conditions are so unstable.

ICO's will further be talked about later in this guide.

Types of Cryptocurrency

Since cryptocurrencies are protected investments and free of political sway, they are considered "digital gold." Peer-to-peer transactions are involved in cryptocurrency exchange. One person can pay another through a computer or mobile device using a downloaded or browser-based app to begin and verify the transaction and transfer the funds. Popularity is being gained by mobile payments which are somewhat similar to cryptocurrency exchange, and are predicted to reach $142 billion in the U.S. by 2019.

Investors and speculators have provided access to a dynamic and fast-growing market by cryptocurrencies, apart from their value as payment mechanisms. Exchanges like Okcoin, Poloniex, and ShapeShift have benefitted greatly from this. In order to fund startups through ICOs, cryptocurrency market has also been used for crowdfunding projects.

The Top Five Cryptocurrencies

Apart from Bitcoin, there were over 1,300 cryptocurrencies on the market by the end of November 2017.

Below we shall discuss the leading five cryptocurrencies by market capitalization:

Bitcoin.

Bitcoin is one of the most commonly used cryptocurrencies to date. Bitcoin was also the first cryptocurrency that was created. Bitcoin has a market capitalization of $180 billion and stands the highest among all the other cryptocurrencies. Bitcoin is considered the gold standard for this industry.

Ethereum.

The second most popular cryptocurrency is Ether. Ethereum blockchain network is powered by a token called Ether. It has a market capitalization of over $18 billion. Ethereum was created back in the year 2015. Ethereum is a Turing-complete programmable currency. This property allows the developers to develop different apps and technologies. This is where Ethereum competes with Bitcoin. Ethereum has the capability of processing complex contracts and programs as well as transactions.

Ripple.

Ripple was created back in the year 2012. Ripple is said to have a market capitalization of $10 Billion. Banks like UBS and Santander have been operating using the Ripple. It is being used because it allows tracking transactions.

Litecoin.

Litecoin is similar to Bitcoin. Litecoin was developed after the creation of Bitcoin. There is a mining algorithm that is used to make the payments fast and process more transactions. These new innovations were a part of the development of Litecoin. It has a market capitalization of $5 billion.

Monero.

Monero is an open-source cryptocurrency that was developed using less transparent CryptoNote Protocol. The algorithm that was used to create this has enhanced security and privacy features over Bitcoin. Being based on an open-source model, consumers have not liked it much as it can help the actions of fraudsters and scammers disguise themselves. Monero has weak growth.

There are a number of cryptocurrencies that exist and these mentioned here are just a tiny part of it. There are signs suggesting that industries have started to consider the development of their own cryptocurrencies. They are doing so as to enable more secure and faster transactions. Let's consider Dentacoin. It has been developed just a while back as the first blockchain platform for the dental industry worldwide.

Understanding Initial Coin Offerings (ICOs)

Initial Coin Offerings (ICOs) has now become widespread. They provide a way to evade the complex and regulated procedure of raising capital from banks or speculating capitalists. ICO is often compared to the crowd funding process as it lacks regularity. In the ICOs, promoters of a startup are pre-sold their cryptocurrency. This is done in exchange for legal tender or other established cryptocurrencies like Bitcoin.

How Does an ICO Work?

The working of an ICO can be better understood by its comparison to the conventional methods. In the conventional methods, the start-up companies raise capital from investors.

Conventionally, a startup company will sell shares to investors in an Initial Public Offering (IPO) transaction.

In the cryptocurrency market, a startup generates coins or tokens. These tokens are used to bargain with the investors in an Initial Coin Offering (ICO) in exchange for legal tender or digital currency.

IPOs deal with investors while ICOs deal with ardent sponsors of their project. This is very similar to crowdfunding.

The cryptocurrency company that comes up with a startup while beginning an ICO campaign makes a plan that sketches the goal that they are trying to achieve in the project. Not just that but it also states all the financial details and how much money they acquire for the project. All the information regarding the sponsors and investors, currency information, and the length of the ICO campaign are mentioned too. When the time for which the campaign was supposed to run ends and the company is not able to increase enough money that is required for the project, the money which the sponsors invested goes back to them.

ICO Success Stories

Ripple is considered to be the first cryptocurrency that raised funding through an ICO. 100 billion XRP tokens were created in order to develop the payment system for Ripple. Those tokens were later sold so that the development of the Ripple platform gets funded.

The most noticeable platform that is involved with ICO funding is Ethereum. Ethereum came up with a smart contract system. In that system, a simple token may be transacted on the Ethereum blockchain. Successful ICO projects were launched through this standard. A few examples of successful ICOs on Ethereum include:

Augur.

In order to fund the development of Augur, the money was raised by selling almost eighty percent of REP tokens. About $5 million of funds were raised which at this time are worth $100 million.

Melonport

Melonport's main objective is to create a platform that would manage the Ethereum-built blockchain assets. MLN tokens were sold for more than 2,000 Bitcoin back in 2016.

Golem.

The main aim behind Golem was to develop a supercomputer. This supercomputer would enable contributors to sell its power. The ICO was limited to 820 million tokens, and the developers received over 10,000 Bitcoin. Today the market share of Golem market share stands at 50,000 Bitcoin.

Singular DTV.

Singular DTV plans to amalgamate with Ethereum, smart contracts, and the production and streaming of videos. The number of Bitcoins that were raised for the purpose was 12,000. These funds were raised through an ICO and now have a worth of 40,000 Bitcoin.

ICONOMI.

ICONOMI is a platform where one can manage and look after their asset. Around 17,000 Bitcoin was collected by the developers who managed to sell eighty-five million ICN tokens in the ICO movement. The market cap today stands over 40,000 Bitcoin.

If you want to see the potential that an ICO has , understand that it virtually knows no bounds. That is because it allows both the companies and individuals to release tokens that can be traded to raise funds. Cryptocurrencies and exchange pose a great influence on the future of financial transactions all across the world. The cryptocurrency model has the capability of security and is able to exceed geographical boundaries. This helps it keep pace with today's digital world but also be a chief motive of innovation.

CHAPTER NO 3

WHAT ARE THE IMPACTS OF CRYPTOCURRENCIES ON THE ECONOMY OF DEVELOPING COUNTRIES?

According to the description of the World Bank, the number of people who live their lives under the poverty line is incredibly high. This shows that the welfare and aids that help the economy grow are distributed disproportionately among different regions and countries. Along with these incidents of economic chaos, civil war, governmental collapse, and plague are developing in regions (Prahalad & Hammond, 2002). Besides, poverty is mainly driven by economic factors which include limited access to financial services (Beck & Demirguc-Kunt, 2006) and high inflation rates (Aisen & Veiga, 2006). Moreover, studies have argued that a low level of trust (Barham, Boadway,

Marchand, & Pestieau, 1995) and corrupt government institutions harm economic development (Olken, 2006).

Crypto currencies could provide a significant benefit by overcoming the lack of social trust and by increasing the access to financial services (Nakamoto, 2008) as they can be considered as a medium to support the growth process in developing countries by increasing financial inclusion, providing better traceability of funds, and to help people to escape poverty (Ammous, 2015).

Before we move on to an overview of the cryptocurrencies and their usage in developing countries, it is important that we understand what are the advantages and disadvantages of the cryptocurrencies delivered for users in comparison to the central bank-issued fiat currencies, identical to the Euro or the US dollar. Not just that but also their deliberate emergence from the fundamental technology. The majority of the cryptocurrencies function using blockchain technology. Basically, blockchain is distributed on different nodes that are a part of the network. The entries of the blockchain are stored in form of blocks.

Bitcoin is the first-ever digital currency that works on an algorithm. The transactions that take place on a decentralized peer-to-peer network are recorded and kept track of. All the participants of the network are able to see and monitor these transactions. The first digital currency that was ever invented is bitcoin. The market cap of bitcoin is over $189 billion.

Satoshi Nakamoto is the inventor of bitcoin. He invented bitcoin in the year 2008 when he published his white paper "Bitcoin: A Peer-to-Peer Electronic Cash System" (Nakamoto, 2008).

Ethereum is an open-source and decentralized platform. Ethereum is based on a blockchain network. It is also a computing platform for smart contracts. Additionally, it backs the improved version of Satoshi Nakamoto's consensus mechanism. Ether is the cryptocurrency that powers the Ethereum blockchain network. Ether is said to be the second-best and popular currency; the first being bitcoin. It has a market cap of $18 billion.

General advantages and disadvantages of crypto currencies

Now, let's discuss the positives and negatives of the cryptocurrencies in comparison to the fiat currencies issued by the central bank and discuss their emergence from the underlying technology. Additionally, we will also draw a comparison to the solutions that exist already. These solutions provide to display the real-world importance of crypto currencies.

One of the top advantages of cryptocurrencies is the combination of creating trusts like accountability and transparency. It lets the users have unrestricted and free exchanges among both parties. The basic technology that runs the blockchain uses consensus mechanisms, hash functions, and public and private key encryption to control transactions. In such scenarios it is not essential for both parties to have trust in each other. Conversely, the user has to trust the network and the underlying blockchain. This is vital so as to secure the blockchain against fraud and attacks.

The currencies that have been issued by the World Bank basically establish trust by third party people. An agent is engaged so as to keep a track of the transactions that happen in a fiat currency. Transactions that are done by intermediaries or third parties are time-consuming and have high costs of the transaction. This results in risking premium for its user.

Decentralized cryptocurrencies have a lot of advantages, one of them being the decentralization of cryptocurrencies that governments cannot cope with. Cryptocurrencies are traded all around the world and cannot be constrained to a specific area. Bitcoin allows low-cost money transfers, principally for those who are looking to transfer lesser amounts of money somewhere internationally like remittance payments (Scott, 2016). Cryptocurrencies lets the transfer of money all around the world without any intermediary. The speed with which the money is transferred increases as there are no third parties that could intervene in the transaction.

However, there are some negative aspects of the broader independent payments which are important to address. One distinguishing aspect is the simplification of the transfer of money that is generated from any illegal activity or by that it makes it easy to transfer money from illegal activities or investing in terror activities. Also, the odds of getting caught or any government interference are next to nil. If you compare conventional money transfers to digital transfers you would conclude that the user in the Bitcoin system is pseudonymous. Opposing a user who uses a bank account, they do not have to get through a "Know Your Customer" (KYC) process. It is a process in which the user has to identify himself in order to have the right of entry to the Bitcoin market.

Furthermore, the decentralization and bitcoin's highly rigid supply schedule lead to instability in its price. The issue of price instability is not faced by bitcoin only, but in fact it is an issue that the majority of the cryptocurrencies have to face. This is a major reason why is storing money difficult and also making contracts in crypto.

The absence of stability in the supply schedule of the Bitcoin symbols in high price volatility.

An additional key feature of cryptocurrencies is that the sustenance of financial presence. That is because they do not require any extraordinary standards of technology. It also allows you to have the right of entry to the internet and a digital device (for example a smartphone) to involve in transactions (Dow Jones Institutional News, 2018).

No institution can impact the supply of crypto-currencies. That is because the supply is well-defined in the core protocol of the crypto-currency (Nakamoto, 2008). Consequently, no state can influence the flow of money, which restricts the power of government.

The forfeiture of power for the government as well as the peril of terror financing has led to a majority of the countries putting a ban on the usage and trading of cryptocurrencies. Indonesia has put a ban on cryptocurrencies due to the same reason. The Central Bank of Indonesia in a press release on the 13th of January 2018 published that it "forbids all payment system operator [...] in Indonesia [...] to process transactions using virtual currency" (Bank Indonesia, 2018, p. 1). This accomplishment represents that some states consider crypto currencies as a danger. This danger overshadows all the advantages that these cryptocurrencies make available for such countries.

Developing countries and poverty

This section conveys a definition of developing countries. Not just that but it will also help in providing an impression that which countries are classified as developing countries. We shall focus on those countries in this section.

A country is said to be a developing country in which the development of the Human Index is low. The industrial base relative to the other countries develops not as much. (O'Sullivan & Sheffrin, 2003). There is no arrangement in which we can say which countries are the developing countries. But the majority of the countries are considered developing countries. These countries have a lot in common and that is something that can define the criteria for the developing countries. These resemblances are scarce supply of food for enormous groups of the population, low per capita income and poverty, absence of educational prospects, a shortage of admittance to quality health care. This is also a reason for the high infant mortality rate and low life expectancy. These factors contribute to a higher ratio of unemployment in these countries and a poor standard of living in general. Additionally, the prevailing assets in developing countries are distributed disproportionately.

"World Economic Situation and Prospects" is a research study that was conducted back in 2018, showing the UN delineate trends and dimensions of the economy of the world. According to this study, all the countries of the world are categorized into three classes. These classes are: developed economies; economies in transition; and developing economies. The arrangement of these groupings anticipates reflecting the economic conditions in these countries (United Nations, 2018). Still, the majority of the countries cannot be categorized as a single category. That is because they have individualities that could divide into multiple categories. The groups that are created by the UN are mutually exclusive. Appendix A contains the list of all the countries that are considered by the UN as developing countries and their classification over geography.

Here, we shall discuss the leading issues that these developing countries have to face. It is not meticulous and there are several other problems that they face like a low educational level and inadequate medical care. One of the major reasons that contribute to poverty is inadequate access to financial services (Beck & Demirguc-Kunt, 2006). A lot of studies show that financial inclusion is vital in order for a country to develop. For example, Honohan (2018) perceived in his study that poverty and access to financial services are associated with each other. Partial access to financial services is a substantial problem on its own. Financial services can be of great help to people because they have the ability to ensure defense at odds with any dearth of money.

Financial intermediaries are of great significance to individuals and companies alike. (Gorodnichenko & Schnitzer, 2013). That is because of the provision of jobs and the ability to release loans. Moreover, it could be made obligatory for the firms to a suboptimal behavior. However, if financial frictions are severe the companies face an unfavorable and harmful situation due to this (Gorodnichenko & Schnitzer, 2013). That is when they are not given a right to financial intermediaries due to which they cannot acquire aid for the innovation, which leads to the creation of competitive disadvantages in comparison to the companies overseas. In addition to that, they cannot abuse the possible complementarities between the innovation and export activities, which results in an increase in the gap in its productivity. (Gorodnichenko & Schnitzer, 2013). Due to this the companies and firms that operate in developing countries are unable to produce the desired revenue. Consequently, backing the local economy less through fewer jobs, lower salaries, and an overall lower tax volume.

Another issue that contributes to the restricted access to financial services for companies and individuals is that they are unable to involve themselves in worldwide trade. This is because they require a bank account with an international transaction identification. Those firms that do not have any bank accounts are not a part of this extensive range of international services and are slowed down in selling their products outside their region (Scott, 2016).

Social trust is another issue that is being faced by developing countries. This is because the economy is improved and developed due to social trust. (Barham, 1995). Social trust and equality, economic equality, and equality of opportunities are connected to each other on a deeper level. The majority of the developing countries struggle with developing social trust. These countries are unable to come out of this situation because they are trapped. The reason for such a situation is that social trust will decrease as long as there is high social inequality. Though, public policies that are able to cure this problem cannot be defined because there is no trust. (Rothstein & Uslaner, 2005).

Corrupt government institutions are another reason why these countries are struggling with development. Corruption results in a loss of welfare as the opportunities are not proportionately distributed among all the citizens. The majority of the people suffer as a consequence of a small number of people who get the advantage of bribes. These people struggle with their government's income. There have been cases of corruption that prevail over the welfares of redistribution programs, such as Olken (2006) presented for Indonesia.

Opportunities through crypto currencies in developing countries

As per the investigation of the economic complications faced by the developing countries, crypto currencies can hasten the process of developing a number of fields. It holds true that innovations are strategic solutions for the catch-up process of developing countries as explained by Chudnovsky and Lopez (2006).

In order to benefit from the advancements that are offered by crypto, people need to have the internet. It means that those people who have the internet can do trading of crypto currencies whereas those who would not be able to do so. Aiming at this reason, it has been seen that the practice of the internet in developing countries has improved intensely over the past decade (Aker & Mbiti, 2010; Tapscott & Tapscott, 2016).

With no cryptocurrencies, the fiat currency needs to be exchanged to those currencies that are widely used like the US Dollar or the Euro. Then it has to be re-converted into the economy which is being used. Subsequently the majority of the time there is no liquid market could be used to exchange the fiat or local currency to the target fiat currency. Optimization of this could be done using crypto- currencies. It would end up making the process faster and cheaper (Ammous, 2015).

Let's consider an example of an Indian worker based in Chicago. This worker can use a local service provider that exchanges US Dollars into Bitcoins in order to transfer money to his family living in India. There the family can deduct Rupees at a local service provider that would change Bitcoins back to Rupees. This process results in making companies like Western Union indisposed. It is still conversely important to have a liquid market that could interchange the Bitcoin with US Dollar and to Rupee, so as to maximize its proficiency. There have been a few startups that were founded for the sole purpose of creating a liquid market for Bitcoins. For instance, BitPesa was founded in Kenya. BitPesa provides liquid markets for some specific currency passageways, e.g. for the direct exchange of Kenyan Shilling to the US Dollar.

Cryptocurrencies can help with getting involved in trading internationally without having a bank account. Bitcoin is a cryptocurrency that can help with the facilitation of businesses and individuals to get involved in international trade on a smaller level. These businesses and individuals can use Bitcoins in exchange for trading goods. This helps in evading traditional e-commerce systems (Scott, 2016), which requires the creation of a bank account.

The financial situation of the developing countries can be improved by using crypto-currencies, which means that cryptocurrencies can help the developing countries to serve as a quasi-bank account. That is because everyone who has access to the internet is able to download a Bitcoin wallet (Honohan, 2008). Bitcoin wallet can be used as a quasi-bank account. This is where people can conduct savings and daily transactions (Scott, 2016).

The high costs that are faced by the transactions can help with increasing the odds for microcredits. This comes at a cost of reducing the transaction costs. If these costs are eradicated there are high chances that international financing could bloom. People from developed countries are able to send money to people in developing countries, all this is possible due to cryptocurrencies as well. As these are small transactions they would be of less money but that could have a lot of impact on the life of a person in a developing country.

At this point in time microfinance transactions like such can cost a lot of money due to the borrowing and then repaying of transactions, which face transaction high fees. This transaction fee is sometimes as much as the money that is being transferred. However, when transaction costs are enormously minimized or disregarded completely, it can make it possible for loans like such to become more popular and such loans could become more widespread (Ammous, 2015).

Additionally, an amalgamation of the crypto currencies and smart contracts can prove to be a defining factor for the solidification of social trust as well as battling corruption via a system that is more transparent. It would become possible for the common citizens to use the data of the cryptocurrencies that are available publically in the blockchain to observe the way in which the funds from the state are utilized. The governments could also benefit from this as they would be able to track and monitor the total expenditure of their money and in what ways they can improve their budget provision. (Schmidt Kai Uwe, 2017).

On the basis of the literature analysis, here we shall focus on the qualitative analysis that is centered around the interviews of experts. Social sites such as Xing and LinkedIn were used to choose an interview partner. The experts that are a part of this differ hugely considering the information and knowledge that they have. They may be representatives of some start-up, a lecturer, or an ambassador at consultancy companies. Another reason that makes them different from one another is their geographical location. As per the region they live and work in, they differ considerably. An overview of this is provided in Appendix B.

Crypto currencies and local fiat currencies

Money has three main functions.

- Money has to be acknowledged and recognized as a medium that can be used in exchange for trading goods and services.
- Money essentially needs to be appropriate as a medium that can store value for saving wealth.
- Money must perform as a unit of account. It can be used to measure and compare the value of goods (Ammous, 2018).

Gold is considered to be the oldest form of money. Cryptocurrencies are matched to the fiat currencies that are issued by the Central Bank as well as gold.

It is of notice that the conventional types of money that are considered to be gold and fiat money achieve all characteristics of money. Cryptocurrencies are actually more appropriate as the exchange medium. That is because the crypto currencies are divisible and can be transferred on a global level. (Ammous, 2018).

One of the major benefits that gold has is that it is the best-collateralized form of money. That is because the medium of exchange has value in contrast to the notes issued by the bank or digital money. Cryptocurrencies are a decentralized form of currency whereas the fiat currency is issued by the Central Bank under the rules of the government thus making it completely centralized. This is a basic difference between fiat currencies and crypto currencies.

The reason which can be regarded for the high instability of the cryptocurrency is that of its decentralization for crypto and the lack of security. So, it destructs the store of the value function and the unit of account function.

One of the major issues that come with crypto-currencies is the instability in their price. Due to this reason during the value transfer, no huge fluctuations in the financial aspect can be seen (Expert 1, 2018). In order to attain a price that is more stable and fluctuates less, the majority of the crypto currencies progress in the direction of a currency that is regulated enough (Expert 5, 2018). Furthermore, a trend in the majority of the crypto currencies is observed to vary from a complete decentralized system to a system that is more centralized. In this centralized system, the players have the power to maximize any possibilities of development. The majority of the cryptocurrencies in use transition from a decentralized system to a centralized system. They do so despite all the odds. This transitioning may lead to a decrease in value. There can be a reduction in the stability of its price through centralization and support provided by economic policy decisions. Cryptocurrencies can develop significantly (Aisen & Veiga, 2006).

Cryptocurrencies have to be restricted to the national boundaries in order to attain political support. This way the governments will have the power and will be able to control the economic parameters so as to retain financial dominion (Expert 6, 2018). However, this would be a way that would completely reduce the pros of the cryptocurrencies and would result in a system that would be centralized.

At this point, we cannot say that the crypto currencies are not a replacement for the currencies that are issued by the government. Additionally, this has given the national authorities a reason to describe crypto currencies as digital assets and not currency. This way it makes it similar to gold. It is also undecided as it needs to be regarded as an asset or as a form of money. It is safe to say that

- Bitcoin is a digital token that can be exchanged by two parties in a transaction.
- If you compare this token to fiat or national currencies, you would see that it has a worth.
- You can use it to exchange for something tangible. It is used occasionally in trivial amounts.

Improvement of financial inclusion in developing countries

A paramount advantage of crypto currencies is the betterment of financial inclusion for the people living in developing countries (Darlington, 2014). The time and cost of the transactions can be reduced by the crypto-currencies. It will act as some sort of bank account that permits people to make their daily transactions as well as save. (Honohan, 2008; Scott, 2016).

Cryptocurrencies can help in making fast and less expensive transactions in comparison to conventional money transfers through bank accounts like the SWIFT process. It can be achieved by eliminating third parties involved, thus resulting in cost reduction and an increase in the speed of transactions (Tapscott & Tapscott, 2016).

If you consider the remittance payments, cryptocurrencies can prove themselves significantly important in the said field as well. Reduced costs of transactions by the usage of crypto currencies will influence microcredits. That is because of the decrease from every transaction in the conversion fee of the banks. The lending process presently is limited to small amounts only, especially for those people who earn a low income. That is because the items that they own are problematic and result in collateralizing with the old-fashioned tools for the money. For instance, if someone uses livestock as collateral, it would be strenuous for the banks to release the loan in this case.

Cryptocurrencies also greatly help in the creation of access to the world market for businesses in order to develop. The customers of these companies pay companies from other countries in cryptocurrencies. The firms need to have an account in a bank with an international identification number. If they do not have this that would be a hindrance in the payments. For instance, you can hire a person in a developing country to do work for you and then you can pay with bitcoins or other cryptocurrencies.

Restrictions and additional prospective capacities of cryptocurrencies

Cryptocurrencies are used to enhance cross-border payments presently because of the minimized costs and less time per transaction. The Libra project that was started by Facebook intends to drastically decrease the cross-border payment fees. This is done so by using blockchain technology. (Groß, Herz, Schiller, 2019). At this time cryptocurrencies are used mostly for cross-border payments. But we can say that in the upcoming future it can be influenced by peer-to-peer lending through a far-reaching market because peer-to-peer lending helps in taking care of the problems caused by liquidity in developing countries. Besides, the collateralization issue could be solved partly by community trust as Expert 6 (2018) has stated.

Crypto currencies also influence the central systems in large organizations or governments. This would greatly help the credibility to a point that the systems used presently cannot help with.

Ethereum can be used helpful in the case of smart contracts because cryptocurrencies are essential to conduct smart contracts as a model that is incentive so people could operate the blockchains and the underlying infrastructure. Ether is majorly used in the services for making payments. On the basis of smart contracts, there are a number of applications that these contracts are an indispensable factor for the future use of cryptocurrencies (Wood, 2014).

Social security is another important application of smart contracts. At this time, numerous layers of bureaucracy handle payments like unemployment benefits and society needs to pay the fee of commission for that. These costs of bureaucracy can be gotten rid of with the help of smart contracts. The social benefit payments could be well-defined in a smart contract.

All of these potential areas require low price instability (Expert 2, 2018; Darlington, 2014). Furthermore, the use of cryptocurrency depends upon the regulatory framework. A relatively sloppy regulatory system would increase price instability. This could prove to be a reason for the mismanagement of the cryptocurrencies for illegal transactions. For example, money laundering. Conversely, a regulatory framework that is too rigid would weaken the advantages of crypto currencies.

In general, crypto currencies can have a substantial influence on developing countries. This can be achieved by maximizing the financial inclusion of individuals and companies. Reduction in the transaction fee and time, cross-border payments can be improved significantly in specific. (Scott, 2016). This is favorable for remittance payments, peer-to-peer lending, and international trade. The battle against the menace of corruption can be fought with technology. This can be done so by having a tracking system that is transparent for the use of funds (Darlington, 2014).

In order to enjoy the benefits that the crypto currencies provide, it is important for the public to adopt and switch to cryptocurrencies. This will help in achieving the basic functions of money. It does not exist presently due to the extreme price instability. The absence of backup and centralization does not support a stable price level (Ammous, 2018). If there are resilient regulation policies and more political support for the cryptocurrencies it would become possible to achieve a high stable price. Cryptocurrencies can however only get political support if the government or central banks get the right to control them. (Jaag & Bach, 2015). However, this would reduce many benefits of the crypto currencies.

Presently, the growth and development of cryptocurrencies are restricted for developing countries in a number of ways. The introduction of cryptocurrencies could help develop the future of developing countries.

CHAPTER NO 4

HOW TO MINE THE CRYPTO

Crypto mining is a way that can help you with receiving cryptocurrency without any investment. You might think, "How are bitcoins and other cryptocurrencies created, and how can you get them without buying them on a crypto exchange?" Initially, the majority of people showed interest in bitcoin and other cryptocurrencies because their prices climbed high. Cryptocurrencies like Bitcoin, Ether, and Dogecoin very much interested people in the early months of 2021 . Crypto exchanges are where you can buy cryptocurrencies as well as trade them. But you can create or mine these tokens on your own PC.

The promise of getting paid with Bitcoin is a major lure for many miners. To be clear, you do not need to be a miner to possess bitcoin tokens. You can buy cryptocurrencies with fiat currency, trade them on an exchange like

Bitstamp with another cryptocurrency (for example, Ethereum or NEO to buy Bitcoin), or earn them by shopping, writing blog posts on platforms that pay users in cryptocurrency, or even setting up interest-earning crypto accounts.

What is Crypto Mining?

Crypto mining is a process in which cryptocurrencies are minted by solving complex mathematical equations and puzzles by using high-power computers. In this process, blocks of data are validated and recording the blocks on a digital ledger. A Digital ledger is referred to as a blockchain. Complex mathematical techniques are used in order to keep the blockchain protected.

Cryptocurrencies use the decentralized way in which help in the distribution and confirmation of transactions that take place. Cryptographic algorithms are used for this purpose. Here there is no central governing authority that looks over the transaction. Also, there is no central ledger.

In order to receive new coins, complex mathematical puzzles are solved in order to help in the verification of digital currencies. All this information is added to the decentralized ledger. The miners receive pay for this whole process. New coins come into circulation due to this mining process.

Working of The Process:

In the process of mining, complex mathematical equations are solved using high-power computers. The one who is able to crack the code has the authority over the transaction. The miners receive a small amount of cryptocurrency for the process of mining. After the miner is able to validate the transaction, it is then added to a public ledger on a blockchain.

How Can You Start Mining?

For your crypto mining, you need a high-power computer. Not just that but you will also need a digital wallet to store your cryptocurrencies and trade them as well. You can become part of a mining pool where you will get more opportunities for earning profit. Mining pools are basically groups for miners where they are able to improve their mining power. The money that is made from the mining process is distributed equally among all those members that are a part of the pool. Through these mining pools, miners get a chance to work together and fight more efficiently.

The algorithm acquires several cryptocurrencies including Bitcoin, Ethereum, and Dogecoin. It guarantees that no single authority becomes so powerful that it starts to run the show. This process done by miners is a crucial part of adding new blocks of transaction data to the blockchain. A fresh block is only added to the blockchain system if a miner appears with a new winning proof-of-work. This occurs after every 10 minutes in the network. Proof-of-work aims to prevent users from printing extra coins they didn't earn, or double-spending.

Mining the bitcoin can be pretty interesting and profitable as well. The majority of the countries are not really into the crypto space because the mining process can be pretty expensive. But there are a few countries that support crypto mining and you will come across great opportunities for your crypto mining. But where do you start? Below is a list of a few countries where you can begin your journey of crypto mining.

But before we dive deeper into that, let's take a look at some of the factors that make a country an excellent spot for mining.

Key drivers for profitable crypto mining operations

There are lots of factors to be considered if you want to try your hand at crypto mining, but here are some of the most common ones you need to be familiar with:

Advanced and specialized mining machines

As we mentioned earlier, crypto miners dig BTC's digital cave to collect fractions of this digital coin. If gold miners use hard rock equipment to dig up nuggets of precious metals; Bitcoin miners assemble their very own mining machines that consist of expensive specialized software and powerful mining rigs.

Crypto miners use this equipment to solve highly complex mathematical problems that are impossible to answer through pen-and-paper and mental approaches. After finding the solution, BTC miners will be rewarded with a newly minted coin and amounts from the transaction fee—but that's a story for another time.

Fast and reliable Internet connection

Besides the costly and powerful equipment, you'll also need a high-speed and stable Internet connection to run the mining operations. Of course, this service also comes with fees which depend upon the part of the world that you live in.

Cheaper electricity costs

It takes up a lot of electricity to mine crypto. This is why some countries do not agree to have activities like such taking place in their border perimeter . If you plan to mine cryptocurrencies, go for a country where the electricity is available at cheaper costs.

Suitable climate conditions

You need to find a country that has an overall cooler climate. That is because the machines that mine crypto run 24/7. A place with cooler climates would help in avoiding the overheating of these machines. It also minimizes the cost of electricity that is consumed by systems that are used for cooling the machines.

Country's economic situation

The economic condition of a country is crucial if you are looking for a country to mine crypto. This can help you in determining how cheap or expensive mining operations for the cryptocurrencies will be. You need to have expensive equipment and skilled professionals to help you monitor them as well. The money that they get in return for their work is calculated on the basis of the country's cost of living. Moreover, you will have overhead expenditures. The place that you would require to place your machines would add up to your total costs.

Government's stance on crypto-related activities

The resources required for the mining of cryptocurrencies affect the environment as well as the power consumption rate of the country. Some authorities remain cynical about digital coins. This is a reason you need to look for those countries that do not find it offensive to mine cryptocurrencies and regard it in a positive way.

Ideal spots to mine cryptocurrencies

There are a few countries that are great to mine crypto and provide you with numerous opportunities alongside as well. The countries that are ideal for mining crypto are mentioned below:

1. Georgia

Georgia is a country that is both cryptocurrency and blockchain technology-friendly. It ranks 109th in October 2020. It has a broadband download speed of 26.80 Megabits per second (Mbps). The price of electricity in Georgia is 0.056 USD per kilowatt-hour (kWh). Also, its temperature is enough to cool the mining machines.

2. Estonia

Estonia ranks 50th in the global ranking list. It has a broadband download speed of up to 74.73 Mbps. The price of electricity in Estonia is 0.174 USD per kWh. Not just that, but there are hundreds of blockchain and crypto companies in the country. It looks at crypto as "value represented in digital form."

3. Canada

Canada ranks 17th on the list. It has the best broadband download rate of 149.35 Mbps. The price of electricity in Canada is 0.174 USD per kWh, Cryptocurrency mining grew in Canada back in 2018 because of the cheap electricity prices and its cold weather. They do allow the usage of digital currencies but they aren't considered legal tender in the country.

4. Norway

Norway closely follows Canada and ranks 18th on the list. Norway has an Internet speed of 146.53 Mbps. In Norway, electricity is generated using hydropower. It snows mostly and the weather is generally cold. It makes it an apt place to cool the equipment used for mining.

Norway neither prohibits nor recognizes crypto. On the contrary, the Norwegian Financial Supervisory Authority (FSA) imposed regulations of money laundering on people who provided virtual currency exchange locally.

5. Kuwait

Kuwait ranks 34th on the list. Kuwait has an internet download speed of up to 110.33 Mbps. The price of electricity in Kuwait is 0.029 USD per kWh. It includes the cost of power, distribution, as well as taxes. Kuwait has problematic regulation issues.

The Ministry of Finance of Kuwait does not recognize cryptocurrencies and the Central Bank of Kuwait prohibits crypto trading for official transactions. In the year 2018, CBK announced issuing an e-currency that they will monitor. If you do consider Kuwait for Bitcoin mining, you have to keep tabs on its regulatory policies.

6. Iceland

The rank that Iceland is on the list is undefined but the positive thing is that Iceland generates its electricity from geothermal resources. That is because Iceland has over 200 volcanoes and abundant hot springs. This provides tons of underground water to be converted for power generation.

7. Switzerland

Switzerland ranks 4th on the list. It has an internet download rate of 186.40 Mbps. This is the fastest among the list of countries mentioned here. The electricity price in Switzerland is 0.228 USD per kWh which is a little costly. In Switzerland, the regulations regarding cryptocurrencies are pretty relaxed. They are categorized as assets or properties. Switzerland is a country that supports crypto.

8. Finland

Finland ranks 35th on the list. It has an internet download rate of 108.84 Mbps. The price of electricity in Finland is 0.183 USD per kWh. Finland has an overall cooler climate.

9. Sweden

Sweden ranks 14th on the list. It has a download speed of 158.73 Mbps. Electricity prices in Sweden are 0.179 USD per kWh. However, these prices vary from area to area and depend upon where you live in.

10. Latvia

Lastly, Latvia ranks 35th on the list. It has a download speed of 115.22 Mbps. The price of electricity in Latvia is 14.2 euro cents per kWh. This cost is said to be the lowest price in Latvia since the year 2014.

Coin Mining in India

In the past few years, mining of the Bitcoin has significantly increased. There are a number of companies that do provide facilities for crypto mining as well as blockchain development in India. However, the mining of Bitcoins in India is an overpriced and risky business. The reason for that is that in India the fight for coins is high and complex. In order to mine Bitcoins successfully, high computational power is essential. For this purpose, high electricity is required and the costs of electricity in India are pretty high.

The electricity price in India ranges between Rs 5.20-8.20 per kilowatt-hour. That is approximately equal to 7-11 cents. Now the total power that is required to mine the coins is 67.29 terawatt-hours a year. This is as per the approximation of the Cambridge Bitcoin Electricity Consumption Index.

Not just that but there is a lack of equipment in India too. In order to have the complete equipment required to mine Bitcoins, it has to be imported from China. This further increases the expenses and reduces profit. Besides that, India also does not have any clear rules for cryptocurrencies. Due to these reasons, India is a risky place to invest in.

The government of India and the central bank have an unclear association with cryptocurrencies. GOI did suggest launching their digital coin in the intervening time. In 2017, India put a ban over the import of ASCI machines. These machines are specifically designed to mine Crypto which pushed Bengaluru-based blockchain technology company AB Nexus to put mining Bitcoin and Ethereum on a halt.

Setting foot on crypto's digital cave

From all this information provided, if we have learned one thing is that there is a lot of research and knowledge required if you are planning to get involved in the field of crypto mining. Explore the information that has been discussed here and see if mining Bitcoin makes financial sense for your crypto needs.

It can be perplexing to fight the right equilibrium between the factors listed above, but just like for the attainment of any goal you need lots of tolerance and persistence for crypto mining. This is one of the initial and chief methods to make profits in the crypto sphere. So, in order to procure sufficient rewards, be certain to prepare the needed resources and let loose your inner crypto genius!

CHAPTER NO 5

WHAT IS BITCOIN?

Bbitcoin was created by Satoshi Nakamoto back in the year 2009. Bitcoin is a digital currency that is decentralized by nature. Pseudonymous Satoshi Nakamoto in his whitepaper talked about Bitcoin. To date, no one knows exactly the person's name who created Bitcoin and its underlying technology. In comparison to the conventional systems of trading, Bitcoins provide a lower fee for the transaction. Contrasting to the currencies issued by the government, digital currencies function by means of decentralized authority.

Cryptography is something that keeps Bitcoin secure and protected. There is a digital ledger that uses blockchain technology to record the history of transactions that take place. This is so to keep the whole process of transactions more transparent and clear. Physical Bitcoins do not exist. Blockchain verifies all those transactions that are added to it. An enormous amount of power of computation is required in order to verify the transactions. This process is known as "mining." Bitcoins are digital currencies and they are not supported and issued by the governments or any banks. Not just that but an individual bitcoin is not considered to be as valuable as a commodity is. Notwithstanding being legal tender in most parts of the world, bitcoin is very popular and has activated the launch of a huge number of other cryptocurrencies. These cryptocurrencies are jointly referred to as "altcoins." Bitcoin is commonly abbreviated as "BTC".

- As per the market capitalization, Bitcoin is considered to be the largest and most popular cryptocurrency. It was launched in 2009.
- Bitcoin uses a decentralized ledger to store, mine, distribute, and trade it. This ledger is referred to as the blockchain.
- The price of Bitcoin and the Bitcoin market is highly unstable. Bitcoin has been seen to have gone through numerous cycles of boom and bust over a very short span of time.
- Bitcoin is the very first digital currency of its kind and has paved the way for other cryptocurrencies.

Understanding Bitcoin

A group of computers also known as nodes or miners is used by the bitcoin system. It runs the code of Bitcoin and is used to store its information on the blockchain. Symbolically speaking, a blockchain can be considered as a collection of blocks. Every block is basically a collection of transactions. Since all the computers that run the blockchain have a similar list of blocks and transactions they can clearly understand these blocks are being filled with new bitcoin transactions so no one is able to rogue the system.

The transactions that happen on the Blockchain network can be viewed by everyone. A fraudster would require to have 51 percent of the computing power that creates the bitcoin to get done with any immoral activity. Bitcoin had about 11,300 full nodes by September 2021. This figure is gradually increasing continuously which makes it difficult to do launch such attacks.

If there were chances that an attack would be launched, those people who are supposed to be a part of the Bitcoin network with their computers or oftentimes referred to as miners would highly likely divert to a new blockchain. This means that the attack that the fraudster was supposed to launch would go to waste.

Two keys are used to store the balance of tokens of Bitcoins. They are the private and the public keys. Private and public keys are strings of numbers and letters that are interconnected via the mathematical encryption algorithm that was used in their creation. The public key which is often compared to a bank account number is basically the address that is available to the world. If anyone wants to send their bitcoins they can use this address for this purpose.

The private key is often compared to an ATM PIN. It is used to authorize the transactions that are happening over the blockchain. You need to keep your private key private and not share it with anyone. Don't mix up the concepts of Bitcoin keys and bitcoin wallet. Bitcoin wallet is a physical or digital device that simplifies the trading of bitcoin and allows users to track ownership of coins. In this context, the word "wallet" may confuse people. Bitcoin is a decentralized currency and does not need to be stored. Rather, it is distributed on a blockchain network.

Peer-to-Peer Technology

Bitcoin was the first cryptocurrency that operated on peer-to-peer (P2P) technology in order to ease instant payments. Those individuals and companies that govern the computing power are referred to as miners and they have the charge of the processing of the transactions that happen on the blockchain.

They are driven by the rewards which are the release of new bitcoin and the transaction fees paid in bitcoin.

The job of these miners is to make sure that the credibility of the bitcoin network is ensured. Not just that but they act as a decentralized authority for this network. New Bitcoins are launched at a rate that is deteriorating periodically. The number of bitcoins that can be mined is 21 million overall. It is being said that there are 18.8 million Bitcoins that exist as of September 2021. There are less than 2.25 million bitcoins left to be mined.

This shows how bitcoins and other cryptocurrencies work differently from fiat currencies. In a banking system that is centralized, the currency is released at a rate that matches the growth in goods. This is a system that is anticipated to maintain the stability of the price. An algorithm enables the decentralized system to release the rate in advance.

Bitcoin Mining

The process through which the bitcoin is circulated is known as Bitcoin mining. In general, we can say that there is a lot of computational power required to solve the puzzles to discover a new block. This block is then made a part of the blockchain network.

Verification of the records of transactions over a network is added by Bitcoin mining. Miners are compensated with some bitcoin; the reward is halved after every 210,000 blocks. In 2009, the block reward was said to be 50 new bitcoins. On the 11th of May in the year 2020, the third halving took place which brought down the reward for each block discovery to 6.25 bitcoins.

Bitcoin can be mined using a wide range of hardware, however, some return higher rewards than others. A few computer chips, called Application-Specific Integrated Circuits (ASIC), and more advanced processing units, like Graphic Processing Units (GPUs), have the ability to return abundant payments. "Mining rigs" are these intricate mining processors.

One bitcoin can be divided up to eight decimal places which are roundabout 100 millionths of one bitcoin. This is the smallest unit and is known as a Satoshi. Bitcoin can be divided further into even more decimal places if the miners participate and agree to it.

Early Timeline of Bitcoin

Aug. 18, 2008

The domain name bitcoin.org is a registered domain. This domain is "WhoisGuard Protected." That means that the identity of a person who is registered in the domain cannot have their information publically available.

Oct. 31, 2008

An individual or a group of people that uses the name Satoshi Nakamoto announces to the Cryptography Mailing list at metzdowd.com: that a new electronic cash system that is peer-to-peer and with no intervention from a third party in works. The whitepaper published on bitcoin.org, which was titled, "Bitcoin: A Peer-to-Peer Electronic Cash System" would become a great charter for the working of bitcoin today.

Jan. 3, 2009

January 3, 2009, is the date when the first bitcoin was mined and "block 0" was generated. Block 0 is known as a "genesis block" and comprises the text: "The Times 03/Jan/2009 Chancellor on brink of second bailout for banks." This was proof of the fact that the mining process was completed successfully on the date mentioned or after that.

Jan. 8, 2009

Cryptography Mailing list announces the very first bitcoin software.

Jan. 9, 2009

9th January 2009 was the date when the first block was mined. It kick-started bitcoin mining in earnest.

Who Is Satoshi Nakamoto?

No person exactly knows who created the first bitcoin categorically. However, the name Satoshi Nakamoto is linked to the person or a group of people who worked on the Bitcoin software and released the original bitcoin whitepaper in 2008. The bitcoin software was released in the year 2009. Ever since then the majority of the people claimed to be Satoshi Nakamoto, but no one knows who Satoshi is convincing. The identity of Satoshi remains unknown to date.

Although it is alluring to believe that Satoshi Nakamoto is an introverted, idealistic genius who invented bitcoin out of thin air as per media, innovations like bitcoin do not normally happen in a vacuum. The majority of the discoveries made by science were based on previous researches, no matter how new and innovative they may seem.

There are predecessors to Bitcoin. Adam Back's "Hashcash" which was invented in the year 1997 followed by Wei Dai's "b-money", Nick Szabo's "bit-gold", and Hal Finney's "Reusable Proof of Work." The white paper that was presented regarding Bitcoin indicates Hashcash and b-money and numerous other works covering quite a few fields of research. Maybe many of the individuals behind the other projects named above have been speculated to have had a role in the invention of bitcoin.

There may be a few reasons as to why the person or the group of people who invented Bitcoin are keeping themselves disguised and not revealing their identity.

- One of the reasons is that of privacy. Bitcoin was able to garner a lot of fame and that fame made it become a worldwide spectacle. This would lead to a lot of media and government attention towards Satoshi Nakamoto.
- Apart from this, another reason may be the potential for bitcoin to create a huge uproar in the present banking and monetary systems. If in case, bitcoin was to achieve public acceptance, this system has the power to exceed the fiat currencies issued by the governments. This move would have led to the governments taking legal action against the person who invented bitcoin.
- One of the other reasons could be that of safety. In the year 2009 alone, the number of bitcoins mined was that of 32,489 blocks. At the reward rate of 50 bitcoins per block the total overhead in 2009 was 1,624,500 bitcoins. This could

lead to the conclusion that in the year 2009, Satoshi and his team were busy mining the bitcoins because they are the ones who own a majority of the bitcoins mined.

- Anyone who owns that much of the bitcoin could get targeted by criminals and thieves. Bitcoin is more like cash and less like stocks. The private keys are needed to approve spending could be printed out and literally kept anywhere.

The one who invented the bitcoin must have taken precautions, but staying anonymous is a great way to reduce publicity.

Special Considerations

Bitcoin as a Form of Payment

Bitcoin can be used to pay for products that are sold and to provide services. Nowadays physical stores also display signs for accepting bitcoin "Bitcoin Accepted Here." These transactions can be controlled with the necessary hardware or address of the wallet. It can be done via QR codes and touch screen apps. An online businesses can also accept bitcoin as a form of payment through credit cards, PayPal, etc.

The first country that officially accepted Bitcoin as a legal tender is El Salvador.

Bitcoin Employment Opportunities

You do have an opportunity to earn money through bitcoin. Those people who are self-employed can generate revenue through the jobs linked to bitcoin. You can earn through bitcoins in several ways like generating an internet service, and to pay for that you can simply add your bitcoin wallet address for payment to the site. A number of websites and job boards allow users who work with digital currencies.

Cryptogrind is a website that helps in bringing together job seekers and potential employers to one place.

Jobs are featured on the Coinality website. Freelance, part-time and full-time job offers. These jobs pay in bitcoin, as well as other cryptocurrencies like Dogecoin and Litecoin.

BitGigs

Bitwage allows you to pick out a percentage of your work paycheck to be converted into bitcoin. That is directed to your bitcoin address.

How to Buy Bitcoin

The majority of people regard that digital currency as the future. The people who use bitcoins consider this facility as a faster, low-fee payment system for transactions all around the globe. You can exchange cryptocurrencies for any kind of fiat currency as it is not supported by any government or a bank. Actually, the exchange rate of the cryptocurrencies against the dollar appeals to prospective investors and traders. It is deliberated as a fundamental and basic reason for the growth of digital currencies like bitcoin. These cryptocurrencies can be used alternatively with national fiat money and traditional commodities like gold.

IRS released a statement in March 2014 and stated that digital currencies like bitcoin, Ether, etc. are to be taxed as property and not as a currency. All types of transactions that use bitcoins or other digital currencies will be taxed. Be it mining of the bitcoin or purchasing from someone else.

Similar to the other assets, the principle of trading high applies to bitcoin as well. Purchasing a bitcoin through an exchange can help you greatly in increasing your currency. However, there are numerous other ways through which you can earn bitcoins.

Risks Associated with Bitcoin Investing

The price value of bitcoin increased dramatically in May 2011 and then for the second time in November 2013. It interested a lot of investors to jump into the bitcoin market even though bitcoin was not designed as an equity investment. Bitcoin is basically a medium that can be used for exchange. But people do not regard it as a medium of exchange and purchase it for its value of the investment.

Nevertheless, there are many risks that come built-in with the bitcoins. Many investor warnings have been delivered by the Securities and Exchange Commission (SEC), the Financial Industry Regulatory Authority (FINRA), the Consumer Financial Protection Bureau (CFPB), and other agencies.

Bitcoins have seen a rise in investments but still, it is a concept that is new. The market is still highly volatile. If you compare the investments in the virtual currency to the conventional investments, you would realize that bitcoin investment is still a risk as there is no history of credibility to back it. These cryptocurrencies are still in the phase of development. CEO of Digital Currency Group, Barry Silbert said that this is a very "highest-risk, highest-return" investment that can be made. Digital Currency Group constructs and capitalizes in bitcoin and blockchain companies.

Regulatory Risk

Bitcoin is an opponent to the currency that is issued by the government. The odds that they may be used for black market transactions, money laundering, illegal activities, or tax evasion are pretty high. Due to these reasons, the governments may end up regulating it, restricting, or banning it for the use and trade of bitcoin. The New York State Department of Financial Services in 2015 settled the regulations for the companies and individuals that are dealing and trading cryptocurrencies. All buy, sell, transfer, or storage of bitcoins to record the identity of customers, shall have a compliance officer, and maintain capital reserves. If any transaction increases more than $10,000. It shall be reported.

As there are no proper regulations regarding bitcoin and other virtual currencies it increases questions over their longevity, liquidity, and universality.

Security Risk

The majority of the owners of bitcoins have not gotten their tokens via mining operations. Instead, these people trade bitcoin and other digital currencies on any of the popular online markets known as bitcoin exchanges or cryptocurrency exchanges.

Bitcoin exchanges are digital and this is the reason that there are so many threats from hackers and malware to this virtual system. Bitcoin theft has increased a lot recently. If the hard drive of any bitcoin owner is hacked by a hacker, he will be able to steal the bitcoin through the private encryption key and then transfer it to some other account. This would result in the owner losing his bitcoin to theft. In order to prevent such frauds and thefts the owner should store the private encryption key over a computer that has no internet connection. Or note down the addresses on a paper and keep it somewhere safe.

Bitcoins are stored in a bitcoin wallet. There are cases recorded where the fraudsters and scammers are able to hack the bitcoin exchanges and then gain access to the digital wallets of the users. For example, in 2014 a really serious occurrence happened in Japan where Mt. Gox was hacked and bitcoins worth millions of dollars got stolen. Afterward, they asked the people who ran the exchange to close down forcefully.

Transactions that happen on a bitcoin network and that involve the trade cannot be reversed. They are permanent. You cannot do anything. No third party can help you get your payment back once you have sent it. It is only possible to get the payment back if it is refunded by the person to whom you sent it. You basically cannot do anything here if you come across a problem. Be very specific and vigilant when you carry out your bitcoin transactions.

Insurance Risk

No federal or government program ensures bitcoin exchanges or bitcoin accounts. In 2019, SFOX states that it will offer bitcoin investors FDIC insurance. However, it would only be provided to the portion of transactions that involves cash. SFOX is one of the leading trading platforms and a prime dealer of cryptocurrencies.

Fraud Risk

There have been cases reported where fraudsters sell people fake bitcoins although verifications of the owners and the registration of the transactions function over the usage of the private key encryption. For example, SEC took legal action over a Ponzi scheme related to bitcoin in July of 2013. Besides that, there have been occurrences of bitcoin price manipulation. Bitcoin price manipulation is also a common type of scam.

Market Risk

The values of bitcoins can swing a lot like in any other investment. In its short life bitcoin has seen a lot of swing in its price. Focusing on the maximized volume of trading on exchanges, it has a high sensitivity to any event that is worthy of the news. As per the CFPB, the price of bitcoin was reduced by 61%

in one day in 2013. In 2014, the price drop of a single day was about 80%.

This digital currency is going to lose its value and become worthless if a majority of the people disregard bitcoin as a currency. There has an assumption that the "bitcoin bubble" had burst when the price decreased from its all-time high during the cryptocurrency increase in late 2017 and early 2018.

If you consider the competition in the market, you will find that it's a lot. There are a number of other cryptocurrencies but bitcoin is on a leading front. That is due to its brand recognition and venture capital money; a technological innovation by introducing an improved virtual coin is always a danger.

Bitcoin reached its highest price of $64,863 on April 14, 2021.

Splits in the Cryptocurrency Community

Ever since bitcoin launched, there are a number of instances in which dissimilarities between factions of miners and developers stimulated splits across-the-board in the community of cryptocurrency. The majority of cases in these groups of bitcoin users and miners have changed the protocol of the bitcoin network itself.

The creation of any new kind of bitcoin with a new name is done through a process called "forking". This split can be a "hard fork." In a hard fork, a new coin shares transaction history with bitcoin up until a decisive split point, at which point a new token is created. Instances of cryptocurrencies that are invented as an outcome of hard forks also include: bitcoin cash which was invented in August of 2017; bitcoin gold which was invented in October of 2017; and bitcoin SV which was invented in November 2017.

A "soft fork" is a change to the protocol that is still compatible with the previous system rules. For example, bitcoin soft forks have added functionality like segregated witness (SegWit).

Why Is Bitcoin Valuable?

Yes, bitcoin is valuable. Its price has increased dramatically from $1 to more than $50,000 in just over a decade. The high value of bitcoin is due to a number of reasons comprising of its scarcity, market demand, and negligible production cost. Bitcoins are impalpable but have a market capitalization of $1 trillion as of 2021.

Is Bitcoin a Scam?

Bitcoins are very real. You may be unable to touch them as they are virtual but they are real. It's been ten years now since the first bitcoin was created. The code that is used to run the system is open-source and downloadable. Anyone can evaluate the code for bugs, or evidence of immoral intent. There are chances that there may be some mishaps but these are not because of the bitcoins but human flaws or due to any third-party application that is being used.

How Many Bitcoins Are There?

The overall Bitcoins to be mined by the year 2140 is going to be 21 million. Presently there are more than 18.8 million (almost 90%) of the total bitcoins have been mined. Researchers say that approximately 20 percent of the mined bitcoins have gone missing. There are a number of reasons for that to happen. Death of the owner; forgetting the private key; or sending the bitcoins to addresses that are no longer in use are a few reasons for those bitcoins being lost.

Should I Capitalize the 'B' in Bitcoin?

While talking about a Bitcoin network, system, or protocol you have to use a capital "B" by convention. When talking about individual bitcoins or as a unit, you have to use a small "b" by convention.

Where Can I Buy Bitcoin?

You can purchase bitcoins from online exchanges. Now the Bitcoin ATMs have also become a common sight all around the world. Kiosks connected by the internet can also help you purchase bitcoins using a credit card or cash. You can also purchase bitcoins directly from someone you know.

CHAPTER NO 6

THE ETHEREUM COIN

Ether is the cryptocurrency that powers the Ethereum blockchain, and Solidity is the programming language of the Ethereum blockchain.

In order to verify and record transactions, a decentralized blockchain network, Ethereum, is used. Applications can be created, published, monetized, and used by users on the Ethereum platform and the payment is made through Ether cryptocurrency. According to insiders, dApps are decentralized applications on the network.

- As of May 2021, Ethereum had the second-highest value in the market and was closely following up to Bitcoin on first.
- Ethereum is said to be an open-source platform that is based on blockchain which creates and shares business,

financial services, and entertainment applications.

- To use dApps Ethereum users pay fees called "gas." They vary with the extent of computational power necessary.
- Ether or ETH is the cryptocurrency running the Ethereum blockchain.
- Ethereum holds the second-highest value in the market and is second only to Bitcoin.

Understanding Ethereum

The main purpose of the creation of Ethereum was to enable developers to build and publish safe and secure smart contracts and distributed applications (dApps), hence enabling users to use them without the dangers of interruption, scam, or interference from a third party.

As defined by Ethereum itself, it is "the world's programmable blockchain." Being a programmable network serving as a safe and secure marketplace for financial services, games, and apps. The payment can be done in Ether, which makes Ethereum unique from Bitcoin.

Ethereum's Founders

A small group of blockchain devotees including Joe Lubin, founder of blockchain applications developer ConsenSys, and Vitalik Buterin launched Ethereum in July of 2015 with Vitalik Buterin being attributed with coming up with the Ethereum concept and now functioning as its CEO and public face. Buterin, born in 1994, is sometimes described as the world's youngest crypto billionaire. Although designed to be used within the Ethereum network, Ether is now an accepted form of payment by some merchants and service vendors including online sites like Overstock, Shopify, and CheapAir.

What Is Ether (ETH)?

Operations on the Ethereum network are eased by the transactional token, Ether. The programs and services that are associated with the Ethereum network need computing power which is not free, and payment is made to execute operations on the network through Ether.

Ether is considered as the cryptocurrency powering the Ethereum blockchain, but it can be accurately referred to as the "fuel" of this network. Each and every transaction in the network is tracked and facilitated by Ether. This is not how an average cryptocurrency works but Ether still has a few properties similar to those of other cryptocurrencies like bitcoin.

Operations on the Ethereum blockchain are powered by the transactional token, Ether.

To change the storage of consumer data like financial records, blockchain development is used by Ethereum technology which is used by third-party internet companies.

As of 2021, Ether is the world's second-largest virtual currency, second only to Bitcoin, by market capitalization.

In 2017 work began to shift the Ethereum network from a proof-of-work (PoW) system to a proof-of-stake (PoS) system, known as Ethereum 2.0, which is not yet fully developed and released.

Understanding Ether (ETH)

To change the storage of consumer data like financial records, blockchain development is used by Ethereum technology which is used by third-party internet companies. A blockchain is an exceptional type of database which stores data in chronologically arranged blocks, formerly used to record bitcoin transactions but presently, it serves as a basis for most major cryptocurrencies.

The Ethereum model intends to create a safer and secure environment for users by protecting their personal data from being hacked. Even though similar to other cryptocurrencies, Ether is a medium of exchange, however, it distinguishes it from other cryptocurrencies by its ability to be used on the Ethereum network to enable the computation of dApps. In contrast to being able to substitute other cryptocurrencies for Ether tokens, you cannot exchange Ether with the other digital currencies in order to provide the computing power that is required for Ethereum transactions.

The building and running of digital, decentralized applications called dapps is supported by Ethereum. You can use Ether tokens to pay for the computational resources mandatory to perform these operations.

Ether acts as a medium to allow payments made by a developer building Ethereum applications who needs to pay in order to host and execute the applications on the Ethereum network, and a user using the application who needs to pay in order to use it.

The number of Ether tokens paid by a developer depends upon network resources used in the creation of an application. This is analogous to how an ineffective engine requires more fuel—and an effective engine consumes less fuel. Data-hungry applications need more Ether for processing transactions. The Ether fee charged to complete the action depends upon the computational power and time needed by an application, for instance, the more computation power and time needed by an application. The fee that is charged in order to complete the transaction is pretty high.

How Is Ether Different from Bitcoin?

As of 2021, Ether is the world's second-largest virtual currency, second only to Bitcoin, by market capitalization. Ethereum blockchain was launched on July 30th,2015 while Bitcoin was first released on Jan. 3, 2009. As opposed to Bitcoin, the number of Ethereum tokens varies and increases constantly according to demand and does not have a specific limit. It is rather limitless hence making the Ethereum blockchain considerably larger than the bitcoin and is expected to continue to overtake bitcoin in the time yet to come.

Another significant difference is the creation of "smart contracts." These can be created when more codes are built into the transactions by contributors to the Ethereum blockchain, hence the network contains executable code. This is not possible in the Bitcoin blockchain since it is simply a ledger of accounts and the data connected to bitcoin network transactions is mainly used for keeping records.

Another difference is the time taken to build a new block which happens in seconds in the Ethereum blockchain while it takes minutes for the bitcoin correspondent to confirm. And most crucially, the biggest comparison is the difference in overall aims and purposes of the networks. Bitcoin offers a safe and protected peer-to-peer decentralized payment system, created as a substitute to traditional currencies. On the other hand, with Ether as a medium to carry out transactions, the Ethereum platform was created with an aim to allow and ease contracts and applications, and it was never planned for it to be an alternate currency or to substitute other mediums of exchange. Rather, the purpose of its creation is the facilitation and monetization of the operations of the Ethereum platform.

Since Ethereum actually supports bitcoin and due to their different reasons for developing, it makes no sense for these two cryptocurrencies to compete with each other from a functional perspective, but they do so for investor dollars because they have both enticed huge amounts of investments from investors.

Plans for Ether

In 2017 work began to shift the Ethereum network from a proof-of-work (PoW) system to a proof-of-stake (PoS) system, known as Ethereum 2.0. The main reason to upgrade to Ethereum 2.0 is to make the underlying network faster and more secure. Advocates of the planned upgrade say that the number of transactions taking place every second increases drastically.

In a PoW system, to authenticate transactions the alleged "miners" compete with each other to solve hard mathematical problems through their computers. In the new PoS system, for the processing of transactions, instead of depending upon miners the Ethereum network will use "stakers" already having some Ether tokens. On Ethereum 2.0 a transaction could be validated when Ether tokens are deposited in a cryptocurrency wallet by a staker using a smart contract - a contract on the Ethereum blockchain that is automatically executed using code.

In a proof-of-stake system, a substantial amount of computational power is not used by stakers because they're randomly chosen and are not competing with other miners. Instead of mining blocks, stakers create blocks upon their selection, and upon rejection they validate those blocks through a process known as "attesting". Participants involved in the process of attesting can receive rewards for proposing new blocks as well as for attesting to ones they've seen.

A game plan for the release of Ether 2.0 was provided on December 2, 2020, by the founder of Ethereum, Vitalik Buterin. The Ethereum 2.0 road map significantly stated that full enactment of the new version would take some time although, the first block of the new Ethereum blockchain was created on December

1, 2020. Despite officially switching to Ethereum 2.0 the network is still contingent upon miners for computing power.

The Ethereum Business

According to Gartner Research, in the category of businesses investing in a blockchain software platform, Ethereum's main rivals, include Bitcoin, Ripple, IBM, IOTA, Microsoft, Blockstream, JP Morgan, and NEO.

At the end of May 2021, the market value of ETH became $2,236.

Disjointedly, Ether is a contender in the highly unstable cryptocurrency market. In May 2021, second, only to Bitcoin, Ethereum was the second-largest cryptocurrency, with its market cap as estimated by Analytics Insight at $500 billion in comparison to that of bitcoin at $1.080 trillion.

Binance Coin, Dogecoin, Cardano, Tether, XRP, Internet Computer, Polkadot, and Bitcoin Cash comprise the other eight on Analytics Insight's list of Top 10.

Ethereum-Based Projects

Quite a few projects are underway to test the claim made by Ethereum regarding its platform's ability to be used to "codify, decentralize, secure, and trade just about anything."

Microsoft is partnering with ConsenSys to offer Ethereum Blockchain as a Service (EBaaS) on the Microsoft Azure cloud. It is envisioned to offer Enterprise clients and developers a cloud-based blockchain developer environment at a single click.

A joint project for creating a network of data centers built on the Ethereum network was announced by Advanced Micro Devices (AMD) and ConsenSys in 2020.

Ethereum's Continuing Evolution

The possibility of using blockchain technology for purposes other than secure buy-and-sell of virtual currency was first seen by the founders of Ethereum. In order to have a secure payment medium for apps built on the Ethereum platform, the cryptocurrency ETH was created.

Since the Ethereum network is safe and protected from hackers, it has now opened up new avenues for the blockchain to be used for storing confidential information such as healthcare records and voting systems. Furthermore, since it depends upon cryptocurrencies, it can be used by programmers for the creation of games and other business applications on the blockchain network.

The Hard Fork

Futile attempts have been made to hack blockchains even though a blockchain may be impenetrable to hacker attacks. In 2016, $50 million worth of Ether had been stolen by a malevolent actor that had been collected for The DAO. It is a project in which a third party has to develop the smart contracts and originate from Ethereum's software platform. A third party developed was held responsible for this efficacious attack.

A "hard fork" was created by the community of Ethereum in order to reverse the robbery which annulled the present blockchain and created a second Ethereum blockchain. The original is known as Ethereum Classic.

Ethereum 2.0

Recorded in May of 2021, the value of Ethereum as a virtual currency was the second-highest in the market, behind only Bitcoin. Back in 2018, the number of ETHs in circulation surpassed 100 million.

Contrasting Bitcoin, an endless number of ETH's can be created.

Presently, Ethereum is being upgraded to Ethereum 2.0, which is anticipated to allow the network to function better and solve the overcrowding glitches that have slowed the network in the past. For instance, in 2017, transactions on the platform were singlehandedly slowed down by a game called Cryptokitties.

Ethereum with its wider aims and ambitions wants to be a platform for all kinds of applications that can store information safely.

Criticisms of Ethereum

Ethereum is subjected to the same criticism as any other cryptocurrency:

Bitcoin price action is imitated by the prices of all cryptocurrencies including Ether. Although this has been obvious for a long time it can barely go unnoticed anymore. For instance, the value of bitcoin fluctuated between about $900 and about $20,000, and then in April 2021 hit a high of around $63k and stayed at around $30k in July 2021. Cryptocurrencies can be subjected to having both highs and lows and remain highly hypothetical and unpredictable.

A large amount of energy is being consumed by every one of these networks. A large amount of computing power is allocated by cryptocurrency miners chiefly to the process of validating transactions. All-encompassing crypto coin mining operations drain fossil energy which is one of the reasons why China is repressing cryptocurrency.

Perhaps, the introduction of Ethereum 2.0 can withstand the criticism that Ethereum had to face due to its fee.

What Is Ethereum in Simple Terms?

Ethereum is an information database, like any other blockchain with Ether being the cryptocurrency used to facilitate transactions on the Ethereum blockchain.

A blockchain organizes data "blocks" of information that are arranged in a chronological "chain." For example, when a transaction is made through an Ether token, it is verified and recorded as an additional block on the Ethereum blockchain. The main reason why a blockchain is often equated to a ledger is this process of sequential verification of transactions.

Apart from storing Ether currency's transaction records, the Ethereum blockchain also permits the creation and marketing of decentralized applications called dApps. Those users want to be benefitted from the comparative absence of dangers that come with storing sensitive information on the internet.

What Is ETH Trading?

Quite a few digital currency trading platforms such as Coinbase, Kraken, Bitstamp, Gemini, Binance, and Bitfinex are used by investors to trade Ether. Investing apps like Robinhood and Gemini also trade in cryptocurrencies.

The prices of cryptocurrencies are extremely unstable and the people trading crypto is making efforts to benefit from this instability. In July of 2021, the value of one ETH, which had surpassed $4,000 in mid of May and was at $231 a year before was fluctuating between $1,800 and $2300.

Is Ethereum Better Than Bitcoin?

In order to provide cryptocurrency support, the bitcoin blockchain was invented. In contrast, the wider-aiming Ethereum blockchain was created as an in-house currency for dApps or any other kind of applications storing information safely.

Regardless of their dissimilarities, Ethereum and Bitcoin are the creators of rival virtual currencies in the investing world. And virtual currencies are nothing but coins without a physical existence. Instead, they are represented by a string of codes that a buyer and seller can exchange at a price agreed.

How Does Ethereum Make Money?

In order to use dApps users are required to pay a fee called "gas" that is conditional on the amount of computational power used.

As recorded by the Ethereum Gas Report at the beginning of 2021, the intermediate fee for gas had crossed $10 per transaction.

How much time is required to mine an Ethereum?

Hashrate, consumption of power, electricity cost, and fees paid to a mining pool associated with the mining operation are some of the factors which affect the time taken to mine Ethereum and obtain mining rewards. The cost-effectiveness and upsurges in mining difficulty, targets, and the general price performance of the crypto market are also affected by these factors. As assessed by the default calculations of an Ethereum mining calculator, it takes 51.8 days to mine ETH.

CHAPTER NO 7

WHAT IS NFTS?

What is a Non-Fungible Token (NFT)?

Non-fungible tokens or NFTs are cryptographic assets on the blockchain. NFTs have unique codes for identification and metadata that differentiate them from each other. You can exchange cryptocurrencies with cryptocurrencies or fiat currencies, but in the case of the NFTs you cannot exchange one with another. That is because NFTs are unique and differ from one another, unlike cryptocurrencies.

WHAT YOU NEED TO KNOW

- NFTs are tokens that are cryptographic. They are unique and reside on a blockchain. You cannot duplicate these tokens.
- NFTs are used to signify real-world items such as artwork and real estate.

- "Tokenizing" these real-world physical assets provide them with an opportunity to be bought, sold, and traded proficiently and resourcefully so as to minimize the possibility of a scam.
- NFTs can also be used to symbolize the identities of people. Not just that but also their properties rights etc.

Every NFT is designed in such a way that it can be used for multiple purposes. For instance, you can use the NFTs to represent physical assets in a digital medium. The physical assets can be anything; be it artwork or real estate. Also, there are no third parties involved in the trading of NFTs. That is because of the underlying technology that powers it. NFTs are based on blockchain networks that simplify and facilitate transactions between two parties. As well as create new markets.

Beeple was able to sell NFTs worth $69 million in March.

This was one of the most expensive transactions that took place via the NFTs. These art pieces are the most expensive art pieces sold ever.

Beeple's first 5,000 days of work were compiled together in a collage and this was the artwork that was sold.

The NFT market generally deals with collectibles such as digital artwork, sports cards, and rarities. NBA Top Shot is the most talked about and famous platform that deals with collecting on-fungible tokenized NBA moments in a digital card form. These digital cards have been sold for millions of dollars. Lately, Twitter CEO Jack Dorsey tweeted a link to a tokenized version of the first tweet ever written where he wrote: "just setting up my twttr." The NFT version has by this time has been bid up to $2.5 million.

Understanding NFTs

Cryptocurrencies are fungible similar to fiat currencies. They can be exchanged with one another. You can trade them easily. An example of this can be that a dollar bill will always be equal to what a dollar bill is worth even if the serial number on both the dollar bills is different. Similarly, you can trade one bitcoin in exchange for another one of equal worth and value. Likewise, one Ether is equal in value to another. This characteristic makes cryptocurrencies an appropriate way to use as a safe and protected medium of transaction in the digital economy.

One NFT can never be equal to another one. That is because the NFTs shift the crypto standard. This ends up making every NFT unique thus making it impossible for two NFTs to be equal. Each NFT is differentiated from another by means of a unique, non-transferable identity. NFTs are basically the symbolization of digital assets. However, you can combine two NFTs and come up with a third NFT. This NFT will be unique as well.

Similar to Bitcoin, NFTs also hold details of ownership so that identification and transfers are stress-free among token holders. For instance, tokens that are demonstrating coffee beans can be classified as fair trade. Artists are capable of signing their artwork by using the stored metadata.

The evolution of the ERC-721 standard gave birth to NFTs. ERC-721 standard was developed by a group of people who were involved in the development of the ERC-20 smart contract. ERC-721 standard describes the minimum interface, ownership details, security, and metadata. All these details are mandatory to be shared so that the exchange and distribution of gaming tokens are made possible. This concept is further taken forward by 1155 standard. This standard minimizes the transaction and storage costs that are a prerequisite for NFTs and batching multiple types of non-fungible tokens into a single contract.

CryptoKitties is regarded as the most famous use case for NFTs. CryptoKitties was launched in November of 2017, and is a digital representation of cats with unique identifications on Ethereum's blockchain. All the "Kitties" are different from one another and each of the Kitty has a different price in Ether. These Kitties give birth to new Kitties. The newly birthed Kitties have different attributes and valuations. CryptoKitties was able to gather a huge fan base that spent $20 million worth of Ether purchasing, feeding, and nurturing them soon after it was launched.

CryptoKitties may be of minor or inconsequential importance but there are several others that have more serious uses in business. For example, NFTs are used in private equity transactions and real estate deals as well. It insinuates empowering many types of tokens in a contract. This is the capability to deliver trust for different types of NFTs, from artwork to real estate, all into a single financial transaction.

Why Are Non-Fungible Tokens Important?

The evolution of the cryptocurrencies concept has given birth to the NFTs. Contemporary finance systems comprise refined trading and loan systems for various types of assets that range from real estate to lending contracts to artwork. NFTs are a step towards the future of facilitating the digital representations of physical assets.

All these little concepts are merged together into smart contracts on a blockchain network. Representation of physical assets digitally or using a unique identification is not a new concept.

Market efficiency is considered to be one of the most important advantages of NFTs. Converting a real-life or a physical asset into a digital asset modernizes processes and eliminates any third parties involved. All the details of the transactions are recorded on a blockchain and this eliminates the need for involving any third party or an intermediary. Owners can directly make a contact with the buyers and finalize a deal. Business deals can also be greatly improved using this concept. For instance, an NFT for a bottle of the vine will make it easier for various members that are a part of the supply chain to interact with it and also be able to track where it originated from, produced, and sold. Ernst & Young has come up with such a solution for their client.

Identity management is greatly enhanced by the use of non-fungible tokens. Let's consider the example of physical passports. You need to display your passports at every entry and exit point. If these passports are converted into the NFTs each with their own unique identification number it would greatly help in simplifying the process of entry and exits for the concerned authorities. NFTs can also be used in digital identity management.

NFTs provide investors to make an investment in a new market and increase their investment opportunities. Suppose there is real estate that is distributed into a number of pieces. Every piece of land has its own distinctive features and types of property. One piece of land might be next to a beach, the other one in an entertainment complex, and another one is a residential district. If you look at the features of all these places you would see that every piece of land is unique and thus will have a different price. Every piece of land will also be represented by a different NFT.

This concept has already been put to use by a platform that goes by the name Decentraland. Decentraland is a virtual reality platform on the Ethereum blockchain. With the rapid evolution that is happening in this field that day is not far when it would become possible to apply the same concept of tokenized pieces of land, differing in value and location, in the physical world.

The internet of assets

One of the few problems that exist on the internet presence can be easily solved by the NFTs and Ethereum. Everything being digitized due to which there feels a need to duplicate the properties of physical items like scarcity, uniqueness, and ownership proof. For example, you cannot sell an iTunes mp3 again after you have bought it, or exchange the loyalty points of the company for the credit of some other platform even if it is in demand. You can't exchange one company's loyalty points for another platform's credit even if there's a market for it.

NFT examples

NFTs are still considered a new concept and the NFT market is a risky one. NFTs are tokens that represent ownership of assets. Below is a list of examples that will help you understand better.

- Unique digital artwork
- Unique sneaker in an unrepeatable fashion line
- Any in-game item
- An essay
- A digital collectible
- A domain name
- A ticket that allows you for an event or a coupon

POAPs (Proof of attendance protocol)

You can ask for entitlement of the POAP NFT if you donate to ethereum.org. You can use these collectibles and show that you were a part of these events. POAPs are used in the

Crypto congregations as a form of a ticket to their events.

ethereum.eth

Ethereum.eth has an alternative domain name that is driven by NFTs. The .org address is handled centrally by a DNS provider. On the other hand, ethereum.eth has been listed on Ethereum using the Ethereum Name Service (ENS).

How do NFTs work?

ERC-20 tokens, such as DAI or LINK are not the same as NFTs. In the DAI and LINK, every token is completely unique and is not divisible. NFTs are a representation of the ownership over a digital asset. It is a unique piece of digital data, that can be tracked by using Ethereum's blockchain. An NFT is a representation of digital objects as a claim of ownership rights of digital or non-digital assets.

For example, an NFT could represent:

- Digital Art
- GIFs
- Collectibles
- Music
- Videos
- Real-World Items:
- Deeds to a car
- Tickets to a real-world event

- Tokenized invoices
- Legal documents
- Signatures

These are not the only ones. In fact, there are a lot of other options as well.

Each NFT is to have one owner at one time. Ownership is handled by uniqueID and metadata. These characteristics are unique to every token and cannot be duplicated. NFTs are minted via smart contracts that allocate ownership as well as manage the transfer of one NFT to another. The creation of an NFT means that the code that is stored in the smart contracts is triggered and executed. It obeys different standards, such as ERC-721. This information becomes a part of the blockchain where the NFT is being managed. Let's walk you through the steps of creation or minting of an NFT below.

- Creation of a new block
- Validation of information
- Recording information into the blockchain

Below are a few special characteristics of an NFT:

- Each token that is created or minted has its own unique identifier that is directly linked to a single Ethereum address.
- They cannot be interchanged openly with other tokens 1:1. For instance, 1 ETH is

the same in worth as another ETH. This is not something that NFTs are concerned about.

- There is an owner of each token and this information can be verified with ease.
- They reside on Ethereum. The buying and selling can be done using any NFT market that is Ethereum-based.

In other words, if you own an NFT:

Proving ownership over an NFT is not a difficult process but a pretty easy one.

You can easily prove that you have ether stored with you in your account as you can prove that you own an NFT.

This denotes that when you purchase an NFT, the unique token that comes with it is then stored in your wallet using your public address.

That unique token is proof that you own the original piece.

The private key you have is proof that you own this authentic piece.

The public key that the creator of content has access to basically certifies the genuineness of a specific digital artifact.

The public key of the creator is basically a permanent part of the token's history. The public key of the creator is a demonstration that the token you now own was created by a so and so.

You can prove your ownership over an NFT in several ways, one of them being messages that are signed to show that you are the owner of the private key on a specific address.

The private key that you have access to is a proof that it is original. This is a way to show that private keys behind a specific address control the NFT.

No individual has the ability to influence it one way or the other.

In case you sell it and receive the royalties of the creator who created the art originally.

If you don't wish it to sell, you can keep it with you for as long as you want to. It will be stored in your wallet on Ethereum safely.

If you are able to mint an NFT:

- Proving that you are the one who created this art would become easy.
- Scarcity can be decided by you.
- Through the royalty scheme, you would be able to earn a percentage of money each time your art is sold.
- Any peer-to-peer market can be used to sell the art without getting any third parties involved.

Scarcity

The scarcity of an NFT is determined by the original creator.

An example below would give you more understanding regarding this topic. Suppose there is a sports event happening. It is only for the organizer of the event to determine how many tickets to sell for the event. Similarly, only the original creator of the asset has the right to determine how many copies he wants to create. Often times the copies are an exact replica of the asset. Other times a number of them are created with a minute difference. In other cases, the creator may want to create an NFT where only one is minted making it one unique and rare collectible.

In the cases explained above each NFT would still have a unique identifier with a single owner. It is the decision of the creator to determine the scarcity of an NFT. It is something that matters. In order to maintain scarcity the creator can create the NFTs completely different from each other. If he wishes, he can also come up with a lot of copies. All this is publically known.

Royalties

If your NFT has a royalty scheme programmed into it, every time your art is sold you will be able to receive your share automatically. This concept is still under works but it is one of the best ones. The ones who own EulerBeats Originals receive an 8% royalty whenever the NFT of their art is re-sold. Platforms such as Foundation and Zora back royalties for their artists.

The whole royalty process is automatic and stress-free. This is so that the artists who created the art originally don't have to go through a hassle and can earn from their art every time it is sold again. Presently, there are a lot of issues in the whole royalties scheme as it is manual. Due to this reason, a majority of the artists are not paid. If you pre-define that you will receive a percentage of money as royalties from the artwork in your NFT, you can keep earning from that art piece.

What are NFTs used for?

Below are a few more famous and widely known use-cases for the NFTs on the Ethereum blockchain network.

Digital content

Gaming items

Domain names

Physical items

Investments and collateral

Maximizing earnings for creators

NFTs are widely used in the capacity of digital art. The reason for that is that the digital art industry has a lot of issues that need to be fixed. The platforms that they sold on would end up swallowing their earning and profits generated thus letting the artists suffer financially.

An artist puts up his work on a social network site and makes money for the platform that sells ads to the followers of the artists. This gives the artist the publicity but exposure and publicity don't make any money for him.

The new creator economy is powered by the NFTs. In the creator economy, the original creators don't pass on the ownership of their content to the platforms.

Every time these artists make a sale, all the revenue generated by the sale goes to them directly. The one with whom he sells his art becomes the new owner. Whenever that owner would sell the art piece further, this would give the artist to have an opportunity to generate some revenue from the sale through the royalties' scheme.

The copy/paste problem

Some people fail to understand the concept of NFTs and claim them to be of no use. The main statement that they pass over such a scenario is that if you can screengrab something and get it for free then why do you have to spend millions of dollars over it?

- If you screenshot or google pictures of a painting by Picasso it doesn't make you the owner of it.
- If you own a real and genuine thing, that means that you won something valuable. The more an item becomes viral, the more in demand it is.
- If you own something that is verified it will definitely have more value.

Boosting gaming potential

NFTs are commonly used in the industry of digital art but they are also very applicable in the game development industry. The developers in the industry use the NFTs to claim ownership over the in-game items. It can help with the in-game economies as well as benefit the players in multiple ways.

You purchase items in almost every game that you play. If the item that you purchase is an NFT, you can end up making money from it outside the game. It can also become a source of profit for you if the item is in demand.

The people who developed the game are the ones who issue the NFTs. This means that they have the right to generate revenue through royalty schemes every time an item is sold again. In such cases, there is a benefit for both the parties (i.e. the players and the developers.)

The items that you have collected in the game are now yours and that the game is not handled by the developers anymore.

If there is no one to handle the game the items would still be yours and you will have complete control over them. Whatever you earn in the game can have a lot of worth and value outside of the game.

Decentraland is a VR game that allows you to purchase NFTs that symbolize the virtual parcels of land that you can use as per your liking.

Physical items

If you compare the tokenization of digital assets and physical assets, you will realize that the tokenization of digital assets is more advanced. There are several projects that are working on the tokenization of real estate, rare fashion items, etc.

In the near future, you will be able to purchase cars, houses, etc. all with the NFTs. This is because the NFTs are basically deeds and you can purchase a physical asset and then get a deed as an NFT in return. With the advancements in technology it will be real soon that your Ethereum wallet would become the key to your car or home. You will be able to unlock your door with cryptographic proof of ownership.

NFTs and DeFi

The NFT space and the DeFi world are beginning to work together closely in several interesting ways.

NFT-backed loans

There are DeFi applications that allow you to borrow money using collateral. For instance, you collateralize 10 ETH so you can scrounge 5000 DAI. If the one who borrowed the money is unable to pay back the DAI the lender will be able to get his payment back. However, not everyone can afford to use crypto as collateral.

Presently, the projects have started the exploration of using the NFTs as collateral. Suppose you purchased a rare CryptoPunk NFT awhile back with a present worth thousands of dollars. That means that if you put this up as collateral you will be able to access a loan with the same ruleset. If you are not able to pay back the DAI, your CryptoPunk will be sent to the lender as collateral. Ultimately, you would be able to see that it can work with anything you tokenize as an NFT.

It is not something that is difficult on Ethereum. That is because both of these share the same underlying technology and infrastructure.

Fractional ownership

NFT creators have the ability to create "shares" for their NFT. Due to this the investors and the fans have a chance to own a share of an NFT and not purchase it as a whole. This is a great way for both the NFT creators and collectors to have more opportunities.

DEX's like Uniswap can help with the trade of fractionalized NFTs. The total price of the NFT can be defined by the price of each fraction.

Fractional NFTs are still a new concept and people are still experimenting with them.

NIFTEX

NFTX

This means that you would be able to own a piece of a Picasso in theory. This is like you would become a shareholder in a Picasso NFT. You would be sharing revenue with every partner that is a part of the NFT. It is highly likely that owning a fraction of an NFT will move you into a decentralized autonomous organization (DAO) for asset management.

The organizations that are powered and backed up by Ethereum permit strangers to become global shareholders of an asset. This would allow them to manage everything securely. They don't have to trust each other necessarily. In these systems, no one is allowed to spend a dime unless approved by every member of the group.

As we have been discussing, this is just the beginning.NFTs, DAOs, and fractionalized tokens are all developing at their own pace. However, their infrastructure has already been built and all these technologies can work in close correspondence to each other as they are all backed up by Ethereum.

CHAPTER NO 8

HOW ETHEREUM IS RELATED TO NFTS

Ethereum and NFTs

E thereum makes it likely for NFTs to work for a number of reasons:

You can prove your ownership over an asset by verifying the history of transactions and token metadata.

No one has the authority to steal and falsify the data and ownership of an asset after the confirmation of a transaction.

You can trade the NFTs without the involvement of a third-party platform or application. This helps in the reduction of overhead expenses.

All those products use the Ethereum blockchain as the underlying technology and due to this reason all these products can comprehend each other well. This is why the NFTs can be transferred easily. You can purchase an NFT on one product and sell it on some other one easily. Ethereum is always running and that makes your token available for trade all the time.

The environmental impact of NFTs

NFTs are becoming more and more famous with every passing day and people have started to pay attention to it as well, but they are also subjected to a lot of inspection over their carbon footprint.

To provide a clear overview:

The increase in the carbon footprints of Ethereum may not be due to the NFTs.

Ethereum may use a strenuous way to secure the funds and assets but its improvement is in the works.

Upon improvement, the carbon footprint of Ethereum will be 99.95% better. This would make it more energy-efficient than many existing industries.

In order to have a better understanding, let's discuss some technical aspects of this:

NFTs do not need to be blamed for it.

Due to the decentralized and secure nature of the Ethereum network, the whole system of NFT is up and running.

Decentralized means that you or anyone who owns something can be verified and proven. You would not require any third party to enforce any rules and regulations. You will be able to do everything on your own. This means that the NFTs can be transferred across markets.

No one can copy and paste your NFT as it is secure.

Ethereum has multiple qualities that make it unique and make it possible for you to own digital assets at a justified price. But it comes at a cost. And that is that blockchains like Bitcoin and Ethereum are energy-intensive at this point in time. That is because it requires a lot of energy to function. If it was easy to rewrite the history of Ethereum history to steal NFTs or cryptocurrency, the system collapses.

The work in minting your NFT

Upon minting the NFTs, the following things would take place.

It is important that you authenticate your asset on the blockchain.

It is mandatory for the owner to update his account balance in order to take account of the asset. Due to this, you will be able to trade and own the asset.

The transactions that are validated and verified are to be added to a block and "immortalized" on the chain.

Every person who is part of the blockchain network has to confirm that the network is correct. This block needs to be confirmed by everyone in the network as "correct". This agreement eliminates the need for any third-party application to become a part of the process because everything is recorded on the blockchain and it shows that you are the owner of the NFT. It is publically available and hence everyone is able to view it. This is a great way for the creators to generate extra revenue.

Miners are in charge to do all these things. These miners are supposed to let everyone on the network know about the real owners of the NFT. Mining is a task that demands a lot of responsibility. That is because if it's not done carefully, anyone could claim that they own the NFT who actually do not. There are numerous encouragements to ensure that the miners do their job with honesty.

Securing your NFT with mining

A lot of computing power is required in order to create new blocks in the blockchain. This makes the mining process pretty difficult. The creation of blocks is done at all times. They are not created when there is a need for it, but every 12 seconds a new block is created and added to the chain.

The Ethereum network is safe and secure and this one quality among many is the reason for the existence of the NFTs. The security of the chain is directly proportional to the number of blocks added to the chain. The addition of more blocks to the chain makes it more secure. If a hacker wants to steal your NFT by altering your data on the block, this would lead to the change in all the other blocks and that would make it possible for those who are running the software to instantly detect it and then stop it from happening.

Computing power has to be used at all times in order to keep the chain secure. The block with no NFTs would still have the same carbon footprint as would the one with NFTs. Other non-NFT transactions will fill the blocks.

At this point, blockchains are energy-intensive.

Every block has its carbon footprint associated with it and this is a problem for the bitcoins too. The NFT's are not the only ones at fault for this.

Mining requires the usage of a huge percentage of renewable energy sources. Not just that but people have also been arguing that NFTs and cryptocurrencies are becoming a problem for the industries due to high carbon footprints. But just because prevailing industries are bad it doesn't mean we shouldn't work hard for the better.

But again we are working for a better future. Making Ethereum more energy efficient is the plan and this has always been the plan.

It's not that we are defending the effects mining has on the environment. As an alternative, we will discuss the changes that can help in making things better:

A greener future

Ever since Ethereum has been developed, the developers are trying hard to focus and make things better. They have been constantly around energy consumption due to the mining process. Developers and researchers alike have put this area under focus and trying their best to solve this issue. And the dream has always been to replace it as soon as they can.

A greener Ethereum: ETH2.0

Ever since the development of Ethereum, researchers and developers have been trying to solve the energy-consumption problems. Due to that, there have been several upgrades in the works. The upgraded ETH2.0 is soon to replace the mining process with staking. Staking will reduce the power of computation as a way to enhance security. This would result in minimizing the carbon footprint by ~99.95%. ETH2.0 would work on stakers committing funds rather than using the power of computation to make the blockchain safe and secure.

The energy cost that Ethereum has will be converted into the cost of running a PC. This is multiplied by the number of nodes that are integrated with the network. Let's suppose there are 10,000 nodes in the network and the cost that is required to run a home computer is approximately 525kWh per year. That would make about 5,250,000kWh each year for the whole network.

This power can be used in comparison to compare ETH2.0 and global services like Visa. 100,000 Visa transactions consume 149kWh of energy. In ETH2.0, the number of transactions would cost around about 17.4kWh of energy or ~11% of the total energy. Here, in this case, we are not bearing in mind the optimizations being worked on in equivalence to ETH2.0, as in rollups. It can be as less 0.1666666667kWh of energy required for 100,000 transactions.

This is important because it helps in the improvement of energy efficiency. Energy efficiency is enhanced as well as the decentralization and security of the Ethereum network are maintained. There may be blockchains that operate on staking but in the case of Ethereum the number of stakers required will be a lot. It can go up to thousands. If you want to make the system more secure, decentralize it as much as possible.

CHAPTER NO 9

WHAT IS BLOCKCHAIN GAMING?

Presently, the gaming industry has been flourishing like never before. The gaming industry is currently valued at $173 billion with very high chances that it will exceed the $300 billion mark in five years' time. The reason for the growth in the gaming industry recently is due to the increase in new players, specifically in mobile games. This has made the gaming industry more sustainable.

The gamers have realized that they have been working more and earning less. They know that they are making big investments but the revenue that is generated is far too less. This has pushed them to discover and explore new ways of earning online. They are coming up with different ways in which they can use their hobby for making more money. This has led to increased growth in the game development sector.

Nowadays the blockchain games are on a rise. These games are providing an opportunity for the players to be the main decision-makers regarding finance, and not the developers. Play-to-earn (P2E) crypto gaming is a trend that should not be overshadowed. Let's have a discussion below on how all this works.

What are blockchain games?

A blockchain is a form of ledger technology that functions as a recording and storing system for information. It is one of the safest and secure technologies because it cannot be hacked nor can any data be modified. The history of the transactions that happen over the network is available publically and everyone can see it. Cryptocurrencies are powered by the Blockchain. Bitcoin and Ethereum are the two leading cryptocurrencies right now and they are creating multiple opportunities for people. Game developers have realized the opportunities it can provide them with.

Normally, in video games or mobile games when a player purchases avatars or other game items like assets, etc. they are not owned by the player. It doesn't matter even if the player is spending real cash to get the assets for the game. That is because these games are built using centralized systems in which the whole control of the dynamics of the game is in the hands of the one who developed it. Everything that happens within the game is controlled by the game developer and the player has zero control. Players do not really own any game assets or accounts, etc. This is not the only issue with this system. Another issue is that the system is not completely transparent and there are high chances for manipulations in the game mechanics.

Fair virtual markets were introduced for the first time using the blockchain. The markets have also become decentralized and players have gained more control over the game mechanics. Blockchain not only works in gaming but also powers virtual economies.

Real ownership: Games that are based on the blockchain make the players be in complete ownership of the in-game assets and other items that they purchase. Assets are typically represented by unique non-fungible tokens (NFTs).

Metaverses and interoperability: The tokens that are used to purchase the in-game assets can also be used to trade on the different platforms on a blockchain.

Fair experience: Games that are based on blockchain technology provide more transparency. That is because blockchain enables the creation of open, distributed, and transparent networks. This has given more control to the players than the gaming companies. Players are the ones who control the game dynamics. Not just that but if they want something to be changed in the game, they can vote in favor of the changes.

Unlimited creativity: Traditional games only ran as long as the developer wanted. That is because those games were run on a centralized server. However, the blockchain-based games run on a decentralized and distributed server and this provides the players to keep playing for as long as they want. It doesn't matter if the developers are involved or not. This helps in expanding the range of assets and boost creativity.

CryptoKitties is the first game of its kind. It is a blockchain-based game and was launched in 2017. In the Cryptokitties game, you can trade and create various kinds of Kitties in the form of NFTs, or non-fungible tokens. This makes each virtual kitten into a unique NFT. The rarer the kitten, the higher the value it holds. One of the most expensive Cryptokittens was sold for $172,000.

This encouraged the virtual market to flourish and expand even more.

Can you earn money from blockchain games?

Blockchain games have two unique features that are:

- Incorporation of cryptocurrencies that can be used for in-platform payments
- The use of NFTs

NFTs allow the users to claim ownership over unique assets. Those assets can be traded with other players within the same game or transferred between platforms as well. You can use these blockchain-based games to make money from the marketplaces in the blockchain games.

Enthusiastic players do have the option of producing physical rewards. It also helps the players earn digital collectibles if that is something that interests them. There is huge potential in this field and this is just the beginning.

People have started to regard this as a full-time job. Others think of it as a prospective investment sector.

How to get started with blockchain games?

If you are looking to get involved in blockchain games, here is what you need to do. First, you need to have a device. It can either be a computer or a mobile phone. After this, you need to choose a cryptocurrency exchange. This platform will help you in converting your money into cryptocurrencies. There are a number of exchanges available but you have to research well and look for the one that is well-suited as per your needs. As the majority of the blockchain-based games are using the Ethereum blockchain, it is advised that you use Ether as it is what powers the Ethereum blockchain network. Next, you need to store your Ether or any other digital currency that you have chosen somewhere safe and secure. You can store your Ether in a digital wallet. There are many digital wallets to choose from. Choose the one that is easy for you to use. Then you have to opt for a game and start playing. Ta-Da! You are good to go!

Top Blockchain Games 2021

If you compare the two types of gaming industries, you would realize that blockchain-based gaming is still a new technology and has not developed much. Although there are a lot of blockchain games available and this number is increasing. These games have racked up millions of players and created a lot of money as a result.

Below are the movers and shakers of the crypto gaming industry:

Axie Infinity

Axie Infinity is considered to be the champion amongst all the crypto games. Axie Infinity has taken NFTs to a new level. It has a Pokemon-like experience and the game is inhabited with Axies. Axies are digital creatures. Each Axie is an individual NFT (or digital creature), with each being an NFT. Players can buy, exchange and breed Axies, and also use them to battle other players or teams in seasonal tournaments. The Axies that are not so commonly available have a very high worth. Awhile ago they were sold for thousands of dollars. In-game tokens like Smooth Love Potion (SLP) and Axie Infinity Shard (AXS) are seeing a huge rise that can go up to 5,700%. This has led to an enormous increase in the virtual economy of the game. This is what differentiates this gaming platform from the others. Players are given an opportunity to cash out in the Axie Infinity game which is a feature that is particular to Axie Infinity only. Also, you can exchange the fiat currencies with the AXS. Axie Infinity proved to be a huge hit in a lot of countries.

Blankos Block Party

Mythical Games marked its entry into the crypto gaming industry with Blankos. In Blankos you are allowed to create your own characters. These characters are created from vinyl toys. These designs are completely customizable. Users are given a chance to collect unique 'Blankos', join quests and be a part of team games ("Block Parties").

Upland

Upland is a blockchain-based game that allows users to buy, sell and trade virtual land mapped to the real world. There are "digital landlords" that build properties and then earn through the UPX coins. The project has mapped the cities of San Francisco and New York and lets you buy virtual properties for sale linked to real-life addresses in both these cities.

Mobox

Mobox is free-to-play. It is a combination of gaming with decentralized finance (DeFi). It runs on the Binance Smart Chain. People who developed Mobox say that this will bring gaming closer to GameFi. It will become such a platform where both the players and investors can access games from different blockchains in one single place. One of the key objectives of this game is to let the players take part in the NFT games with ease and be able to make money just by playing the game. Processing of transactions, staking and governance can be done through the MBOX tokens. This would greatly help in encouraging development and allocating the resources inside the MOBOX ecosystem. Presently the number of games that are available on the platform is three and more will be released soon.

Lightnite

Lightnite was developed by the same team that worked on the Bitcoin arcade game portal "Satoshi's Games." Lightnite is a battle royale game. In this game the players are given prizes in the form of bitcoin if they shoot other players in the game. Lightnite is a multiple-player game. The more people you shoot in the game the more bitcoins you yield as a prize. Players who are shot are punished by losing their bitcoins. The in-game assets can be tokenized and used as NFTs. These can be sold and traded via an NFT marketplace.

Gods Unchained

Magic was the basic motivation behind this game. Gods Unchained is a blockchain-based game. In this game, the users create decks and then play cards among themselves in order to fight the opponents. Being a blockchain-based game, it allows the users to have complete control over their cards. Not just that but in case a player wins, he can sell and trade the cards as well. The cards that are rare in the game are worth thousands of dollars. There is a free-to-play option for those players who may be confused and not want to go all-in.

CryptoKitties

CryptoKitties is the first game of its kind. It is a blockchain-based game and was launched in 2017. In the Cryptokitties game, you can trade and create various kinds of Kitties in the form of NFTs, or non-fungible tokens. This makes each virtual kitten into a unique NFT. The rarer the kitten, the higher the value it holds. One of the most expensive Cryptokittens was sold for $172,000. The user interface of the CryptoKitties game is pretty simple. It is highly addictive and you won't be able to stop once you start playing it! The problem with CryptoKitties is that breeding and purchasing the Kittens usually costs a lot and can you may have to spend a lot of Ether on that.

Splinterlands

Splinterlands is reminiscent of Pokémon, Yugi-oh, and Magic The Gathering. Splinterlands is one of the most successful examples of a card fighter experience with a player-driven economy. All the cards that are involved in the game are of real value. You can trade them if you want to. Splinterland is a relatively easy game. There is an in-game shop from which users can purchase cards. There are multiple updates for Splinterland and due to these expansions a huge number of people have started to join the game as players. Rare cards are more tempting for the players to join the game.

Sandbox

Sandbox is a metaverse in which a player can own land, build, play, and take part in virtual experiences. The Sandbox is a video game where players use the Ethereum blockchain to monetize their experiences. SAND is a cryptocurrency that is used as a usage fee for the game. It is a kind of utility token. A web-based marketplace that lets users upload, publish and sell creations made in VoxEdit, a 3D voxel modeling package, as NFTs. Game mechanics can be easily changed using the scripted behaviors after the creations are published and purchased. An editor can be used to place on land parcels. These in turn make changes to the game mechanics.

Cryptopop

Cryptopop is a game that is similar to the widely popular game Candy Crush. Players can receive points by matching the symbols of cryptocurrencies. The more candies you match; the more points you receive. There is an in-game market through which players can earn extra money. Not just that but they can also trade with other players who are playing the game. Ether and Popcorn are the currencies that are used in Cryptopop.

Illuvium

Illuvium has not been released yet. The developers are planning to release it soon. Illuvium has become the talk of the town because as per the developers it is supposed to be the first game of its kind. It will be the first AAA game on the Ethereum network. It has caused quite an uproar and it is justified because this open-world fantasy battle game is built on the Ethereum blockchain and is populated by creatures called Illuvial. Each Illuvial has its own unique capabilities, classes, and categories. These Illuvials can be captured and then used in the fight in order to battle other Illuvials. Players can capture them and have them battle other Illuvials. These creatures can be sold in exchange for ILV tokens. The profits that are earned go all to the players and stakeholders of the game.

Why should you try crypto games?

The concept of the players earning from video games is not a new concept. Games that are based on the blockchain network provide an opportunity for the game developers as well as the players to earn from playing these video games. You will earn money through these in-game assets as well as you can claim ownership of these assets in the game. It also makes you a stakeholder by backing a community-driven ecosystem (DAO). The advantages that these blockchain-based games have were not possible in the conventional gaming methods that worked on centralized platforms.

CHAPTER NO 10

EVERYTHING ABOUT SMART CONTRACTS

Programs that are stored on the blockchain are known as smart contracts. These programs have a set of instructions that execute only when the condition is met. Smart contracts are used in the automation of the execution of a contract. In smart contracts, all the members that are a part of the contract are certain of the outcome without any third party getting involved. Whenever the conditions are met, it triggers an automatic workflow.

How smart contracts work

Smart contracts work using the "if/when...then..." statements that are written into code on a blockchain. Whenever a certain condition is met, the lines of code that are stored in the blockchain get executed. These actions could be anything from releasing funds to the appropriate parties, registering a vehicle, sending notifications, to issuing a ticket. When the transaction is over, the subsequent updates are made into the blockchain. This makes the transaction impossible to alter. Only those people who have the permissions will be eligible to view the results that are generated.

There are many conditions in a smart contract. The number of these conditions can be as per the needs of the user or how many satisfies him that the task will be completed reasonably. Terms and conditions are supposed to be provided by the participants. He should determine how transactions and their data are represented on the blockchain, agree on the "if/when...then..." rules that administer those transactions, explore all possible exceptions, and define a framework for solving any disagreements.

The developer would program the smart contracts as per the rules defined by the participants.

Benefits of smart contracts

Speed, efficiency, and accuracy

The execution of the smart contract begins at the very instant when a set of conditions are met. Smart contracts are digital and automated. Smart contracts do not involve any paperwork. In the manual system, there are a number of errors that may take time to resolve but as smart contracts are automatic, there is no time wasted in this regard.

Trust and transparency

Due to no intermediary being part of the whole process and sharing of the encrypted records amongst all the participants, it makes it difficult to make any changes for personal gains.

Security

Smart contracts are secure. That is because the transaction records that are stored on the blockchain are in an encrypted form makes it almost impossible to hack. Every record on the blockchain is linked to the previous and subsequent records on a distributed ledger. It would take the hackers to hack the entire change and make the changes.

Savings

Smart contracts eliminate intermediaries to handle transactions and, by extension, their associated time delays and fees.

Applications of smart contracts

Safeguarding the efficacy of medications

Sonoco and IBM are working on increasing the supply chain transparency so that it is easier to transport lifesaving medications. Powered by IBM Blockchain Transparent Supply, Pharma Portal is a blockchain-based platform that tracks temperature-controlled pharmaceuticals via the supply chain to provide trusted, reliable and accurate data to all the concerned parties.

Increasing trust in retailer-supplier relationships

Smart contracts are used to solve clashes with the vendors. The Home Depot uses smart contracts for this purpose. They are able to do so via real-time communication and improved visibility into the supply chain. This helps them in maintaining a good relationship with the suppliers. This is a great way for innovation.

Making international trade faster and more efficient

We.trade uses identical and standardized rules and simplified trading options so they are able to minimize the friction and risk while the whole trading process is easy. This helps in expanding the trade opportunities for the companies and banks that are a part of the network.

Smart contracts are automatic and execute on their own whenever certain conditions are met. Business automation applications run on a decentralized network such as blockchain.

Smart contracts are a great way to eliminate any extra expenses. This makes smart contracts one of the most interesting and attractive features of blockchain technology. Blockchain is a sort of database that helps with verifying transactions that may have happened. Smart contracts execute pre-determined conditions. The smart contract can be regarded as a computer executing on "if/then," or conditional, programming.

The whole process of smart contracts is pretty simple. What happens is that the conditions that are required to be met are coded in the smart contracts. When those conditions are met, the lines of code execute and goods arrive in a port. The parties that are involved in the transaction settle for an exchange in cryptocurrency. Transferring fiat money or the receipt of a shipment of goods that allows them to continue on their journey can be automated as well. The underlying technology used is the blockchain ledger that is used to store the state of the smart contract.

Understanding tokens and smart contracts

Let's consider an example of an insurance company. The insurance company could use smart contracts to automate the release of claim money based on events such as large-scale floods, hurricanes, or droughts. When the shipment reaches a port of entry and IoT sensors installed inside the container confirm the contents have not been opened and are stored properly throughout the journey, a bill of lading is issued automatically.

You can use smart contracts to transfer the cryptocurrency and the digital tokens through the network. ERC-20 and ERC-721 are the two tokens that are used in the blockchain network. Alongside tokens, they are smart contracts as well.

But this does not imply that all the tokens are smart contracts. As per Martha Bennet, you can have smart contracts running on Ethereum that trigger an action based on a condition without an ERC-20 or ERC-721 token involved." Martha Bennet is the principal analyst at Forrester Research.

Transfer of cryptocurrencies from buyer to seller can also be administered by smart contracts. Once payment is verified, bitcoin can be traded.

You see that majority of the blockchain networks do not use the tokens. All the rules for it are stated in the smart contract, from the token allocation process to conditions set for transfer.

"That still doesn't mean the token is the smart contract - it all depends on how the token has been constructed." Bennett said. "And tokens don't have to be about economic value; a token can simply be something you hold that gives you the right to vote on a decision; casting the token means you've voted, and can't vote on this decision again – no economic value associated."

How smart contracts mimic business rules

Smart contracts are basic rules for businesses that are coded into software. On the contrary, to the name smart contracts, they are not smart and not contracts legally as well.

"People often ask what makes smart contracts different from business rules automation software or stored procedures. The answer is that conceptually, the principle is the same; but smart contracts can support automating processes that stretch across corporate boundaries, involving multiple organizations; existing ways of automating business rules can't do that," explained Bennet.

She further clarified, "In other words, they're code that does what it's been programmed to do. If the business rules...have been defined badly and/or the programmer doesn't do a good job, the result is going to be a mess." Bennett said. "And, even if designed and programmed correctly, a smart contract isn't smart – it just functions as designed."

"Translating business rules into code doesn't automatically turn the result into a legally enforceable agreement between the parties involved (which is what a contract actually is). Although there are some initiatives aimed at making smart contracts automatically legally binding, that path, at least for now fraught with difficulty and risk." Bennett said. The reason for this that there's no decided standard definition of what is a smart contract.

"And what happens if the software has bugs and yields bad results? Is the resulting loss now also legally binding?" she further added.

The importance of good data, and 'oracles' in smart contracts

It is very important to ensure the quality of programming for smart contracts. That is because a smart contract would work better if the rules and regulations are properly coded to automate processes. The data that is to be added to the smart contract needs to be done vigilantly. Due diligence is required because once you code the smart contract you would not be able to make any changes to it. Not even the programmer could help you in this regard.

The smart contract would be unable to work properly if the data that is coded into it is wrong or false.

Blockchains have the data that triggers the execution of smart contracts. These happen from using external sources like data feeds and APIs. Direct fetching of the data cannot be done via a blockchain. The real-time data feeds for blockchains are called oracles. These oracles are kind of an intermediate layer that is present between the data and contract.

Oracle may be:

- Software-based
- Hardware-Based

A hardware-based oracle, for example, might be an RFID sensor in a cargo container that sends the location data to smart contract parties. A software oracle could be an application that feeds information through an API about a securities exchange such as changing interest rates or fluctuating stock prices.

Smart contracts run on one node. Unlike the blockchain, they are not decentralized and are able to run on a number of computers. The nodes that are part of the blockchain do not know the working of a smart contract. The number of companies that are involved has to trust the oracle for the information that is being added into the smart contract.

No company can know what is happening in the smart contract if your company is part of the blockchain network. There is absolutely no sustainability. In any case, you would have to trust the company that is working on the server that the information added to the blockchain is correct. The company works on the server on which lives the oracle and smart contracts.

"You have to go to one source, one table, one oracle for that data. There are no standard processes to verify the data is what it says it is and it's coming in properly. It's a central point of failure." explained Gartner Vice President of Research Avivah Litan.

"It's not mature yet." Litan continued. "I've talked to companies participating in a consortium and asked them how do you know what the smart contract is doing and they say they don't. If you have a contract running your life, wouldn't you want to know what it's doing?"

Potential problems with smart contract data

Oracles usually transmit data from a single source, thus the data is not to be trusted completely. As per Sergey Nazarov, CEO of Chainlink, an oracle start-up that uses multiple external sources of oracle data. Nazarov, in a white paper, talked about this and said that there are chances that the data may be "benignly or maliciously corrupted due to faulty websites, cheating service providers, or honest mistakes."

Development partnerships have occurred between the internet and financial services by Chainlink. These comprise of Google and the Society for Worldwide Interbank Financial Telecommunication (SWIFT). SWIFT runs one of the world's largest clearing and settlement networks.

Nazarov says the functionality of the smart contracts presently can be tricky. That is because one party may perform a task but the other party may choose not to pay. This would instigate a legal battle between both parties.

"Those contracts are not rigorously enforceable; they can't be enforced by technology the way a smart contract can." Nazarov said. "A smart contract is deterministic; it can absolutely be enforced as long as the events related to its contractual clauses happen".

"Smart contracts are contingent on events; they're contingent on market events, in insurance they're contingent on IoT data from cars, factories, or other equipment." Nazarov continued. "In trade finance, they're contingent on shipping data."

Let's consider an example, Chainlink made a smart contract for a media company. The company did not pay the search engine optimization (SEO) firm that was hired until news article URLs reached and then sustained search engine rankings for some time.

"That payment wasn't held by our client or the search engine optimization firm," Nazarov said. "It was held by this new technology (blockchain and the smart contract) that will programmatically enforce the contract as it was written. That's the fundamental difference."

"The emergence of new technologies and improvements in programming tools has made the creation of smart contracts easy in comparison to the past. These programming tools have helped in moving away from the underlying complexity of smart contract scripting languages, essentially enabling business people to pull together the basics of a smart contract." Bennett said.

"We're even beginning to see tools that allow business people to pull together the basics of a smart contract," said Bennett. "That's only the beginning, though, as some companies have already discovered it can be a challenge to ensure that every network participant runs the same version of a smart contract."

Edge computing, IoT and the future of smart contracts

The use of IoT devices has increased multiple folds. In the coming years there will be a high surge of IoT-connected devices that can shoot better use of smart contracts. Regarding the Juniper research, it is approximated that approximately 46 billion industrial and enterprise devices connected in 2023 will depend on edge computing. This would result in making standardization and deployment critical.

Eliminating any third party involved has increased the speed of the exchange of data and this could prove to be a standard for improving the exchange speed between the IoT devices. This could be made possible due to smart contracts. The server or cloud service that acts as the central communication spoke for requests and other traffic among IoT devices on a network is said to be a middle man or a third party.

Member of the Institute of Electrical and Electronics Engineers (IEEE), Mario Milicevic said, "Fundamentally, the idea is you don't have a central agent – no one approving and validating every single transaction. Instead, you have distributed nodes that participate invalidating every transaction in the network."

The time that is required for the IoT devices to exchange information and processing time has been significantly reduced due to the Blockchain ledgers.

"It could be in an automotive manufacturing plant. As soon as a certain part arrives, that part then communicates to other nodes at that destination which would agree that part arrived and communicate it to the entire network. The new node would then be allowed to begin doing its work," Milicevic said.

Blockchain experts from IEEE are of the view that merging together blockchain and IoT could prove to be a game-changer and would actually transform vertical industries.

At this point in time, the financial services and insurance companies are leading presently in the blockchain development and deployment, transportation, government, and utility sectors are now also becoming a part of it. There has been a huge focus on the opportunities due to process efficiency, supply chain, and logistics. These will most likely merge with the smart contracts in the coming years and that will make the whole process more universal.

CHAPTER NO 11

WHAT ARE THE BLOCKCHAIN'S CONSENSUS PROTOCOLS?

Satoshi Nakamoto was the first one to speak about bitcoin or cryptocurrencies in the white paper that he presented in 2009. After this, there were a number of other cryptocurrencies launched as well. All these digital currencies have been doing really well. Cryptocurrencies like bitcoin and Ethereum, etc. are decentralized currencies. Bitcoin has elevated blockchain technology to new highs and caught the attention of people ever since it was launched. Afterward, many cryptocurrencies and projects based on the blockchain sprung up. This resulted in blockchain becoming the talk of the town with everyone showing great interest. However, the technology on which blockchain works is not a new one. Blockchain is simply an amalgamation of cryptography, distributed system technology, peer-to-peer networking technology, and other famous and widely used technologies. Cryptocurrencies also use the blockchain network. The blockchain network acts as a secure framework for these cryptocurrencies. Due to this nobody can modify or falsify any information related to the transactions. All the nodes that are a part of the network are anonymously added to the network. This makes blockchain technology one of the widely used technologies in various fields, (e.g., financial field, medical systems, supply chain, and Internet of Things (IoT).)

Application of the blockchain technology could however result in a lot of challenges that need to be addressed. Designing the appropriate consensus protocol is also a problem. The consensus of blockchain is that all nodes maintain the same distributed ledger. Centralized servers in the past made the consensus protocol easy to implement. That is because all the nodes would be connected to one central server. This could however prove to be a huge problem in distributed systems because every node is both a host and a server. Both these nodes are supposed to exchange information with the other nodes to reach a consensus. The process of consensus could be disrupted due to offline and malicious nodes. Hence, an excellent consensus protocol can tolerate the occurrence of these phenomena and minimize the damage done so the result of the consensus remains unaffected. Another thing one needs to take care of in the consensus protocol is that you need to choose the consensus protocol that is suitable for the type of blockchain being used. Every type of blockchain has different application scenarios.

Here we shall discuss a few consensus protocols blockchain and investigate their performance and application scenarios.

Main consensus protocols

If you consider the distributed systems, you would conclude that there can never be a perfect consensus protocol for them. The consensus protocol needs to make a trade-off amongst consistency, availability, and partition fault tolerance (CAP). Here we shall discuss some popular blockchain consensus protocols that can effectively address the "Byzantine Generals Problem" in great detail.

PoW (Proof of Work): PoW protocol is implemented by Bitcoin, Ethereum, etc. PoW chooses one node to create a new block in every round of consensus by competition of the computational power. A cryptographic puzzle is required to be solved in order to become a part of the competition. A new block can be created by the node that gets done with the puzzle first. The flow of the creation of blocks in PoW is presented . The puzzle is very complex and difficult to solve. Nodes need to keep correcting the value of nonce to get the answer correct. This adjustment of the value requires huge computational power. PoW belongs to the probabilistic-finality consensus protocols since it guarantees eventual consistency.

PoS (Proof of Stake): In PoS, the creation of the new blocks by the nodes depends upon the held stake rather than the computational power. Despite this , the nodes are still required to solve a SHA256 puzzle: **SHA256(timestamp,previous hash...)<target×coin.** The basic difference between the PoW and PoS is that PoS solves this puzzle by adjusting the value of the nonce and PoS solves this puzzle by the amount of stake (coins). From this, we can say that the PoS is an energy-saving consensus protocol. It does not require a lot of computational power to reach a fair consensus. The flow of PoS is shown . Similar to that of PoW, PoS is also a probabilistic-finality consensus protocol. The very first cryptocurrency that applied the PoS to the blockchain is PPcoin. Another thing that helps in solving the puzzle along with the stake is the age of the coin. For example, if you have 10 coins for a total of 20 days, then the age of your coin is said to be 200. Whenever a node

is able to create a new block, the age of the coin would drop down to zero. Additionally, Ethereum is planning its transition from PoW to PoS.

DPoS (Delegated Proof of Stake): DPoS operates on the principle that allows the nodes that hold a stake to vote to elect block verifiers or block creators. This provides the stakeholders with the right of creating blocks for the delegates they support instead of creating blocks on their own. This results in bringing their computational power consumption down to 0. The delegates are supposed to create blocks whenever it is their turn. If they are unable to do so, they would be eliminated and then selection of new nodes would take place by the stakeholders. In order to reach a fair consensus, DPoS uses the votes of the shareholders. In comparison to the PoW and PoS, DPoS is low in cost and has high inefficiency. Some of the cryptocurrencies are also approving DPoS such as BitShares, EOS, etc. The new version of EOS has turned DPoS into BFT-DPoS (Byzantine Fault Tolerance-DPoS).

PBFT(Practical Byzantine fault Tolerance):

PBFT is a Byzantine Fault Tolerance protocol with low algorithm complexity and high practicality in distributed systems. There are five phases in PBFT: request, pre-prepare, prepare, commit, and reply. The primary node sends the message that is sent by the client to the other three nodes in the network. If node 3 crashes, a single message has to go through all five phases to reach a consensus amongst these three nodes. In order to complete a successful round of consensus, these nodes reply back to the client. In every round of the consensus, PBFT makes sure that all the nodes that are involved preserve a mutual state and take stable action. The Absolute-finality of the PBFT protocol is that all the nodes achieve a mutual state and the protocol achieved what it was aimed for. Stellar is a new protocol. It is basically an improved version of the PBFT. Stellar adopts FBA (Federated Byzantine Agreement) protocol, in which nodes have the right to opt for the federation they trust to carry out the consensus process.

Ripple: Ripple is an open-source payment protocol. Transactions are commenced by clients and broadcast all over the network through tracking nodes in Ripple. In Ripple, the consensus process is done via validating nodes. Every validating node has a list known as UNL (Unique Node List). All the nodes in the UNL have the right to vote for those nodes that they are in support of. Each validating node directs its own transactions fixed as a proposal to other validating nodes. When the proposal is received by the other validating node, it will look at each transaction in the proposal. In its local transaction set, if there exists the same transaction, the proposal would receive a single vote. The transaction would be able to enter into the other round if it was able to receive more than 50% of the votes. This number would increase for every other round. Those transactions that receive more than 80% votes will be added to the ledger. This makes Ripple is a consensus protocol that is able to achieve absolute finality.

Fault tolerance

PoW, PoS, and DPoS are probabilistic-finality protocols. The attackers would be required to gather huge computational power in order to create a long private chain to swap with a chain that is valid. In Bitcoin, an attacker would require to have a 50% fraction of the computational power to create a longer private chain. If the computational power of the attacker's fraction is more than or equal to 50%, it would destabilize the blockchain network. Like PoW, PoS and DPoS can only allow the existence of the stakeholder with less than 50% of the held stake. In PBFT, if there are a total of 3f+1 nodes in the network. The number of normal nodes must surpass 2f+1. This means that malicious or crashed nodes must be less than f. The fault tolerance of PBFT is 1/3. The fault tolerance of Ripple is only 20%, (i.e., Ripple can tolerate Byzantine problems in 20% of nodes in the entire network without affecting the precise result of consensus.)

Limitation

PoW may consume the highest computational power among these consensus protocols, and the transaction throughput per second (TPS) of Bitcoin adopting PoW is only 3–7. This speed limits the application view of PoW in real payment. PoS and DPoS also have similar inadequacies but they can help with bringing down the consumption of computational power. PBFT needs every node interacting with other nodes to exchange messages in each round of the consensus. PBFT has the highest requirements of performance for the network. There is no secrecy in PBFT as all the nodes have their identity revealed in the consensus. It requires a few seconds to finish a round of consensus Ripple. It is appropriate for the actual payment scenario. Ripple is organized and handled by some organizations which does not fulfill the decentralization nature of blockchain.

Scalability

PoW, PoS, and DPoS are relatively scalable. Although the TPS of them is not very high, there are some ways that can help improve the scalability. For example, in order to enhance its scalability bitcoin adopted a lightning network to deliver an off-chain payment. Ethereum chose sharding technology and Plasma. Both of these are layer 1 and layer 2 scaling solutions, respectively. As PBFT is appropriate for a network with fewer nodes and high performance the scalability of PBFT is inadequate. For networks that work on a large scale, Ripple can prove to be useful. TPS of Ripple is over 1500, hence Ripple has strong scalability.

Scenarios

The blockchain systems presently used can be divided into three categories. In a public blockchain, everybody has the right to be a part of the process and can view the distributed ledger. PoW, PoS, and DPoS can be applied to the public blockchain. Private blockchain and consortium blockchain fit the permissioned blockchain. Only those nodes can take part in the consensus process who have permissions. The identity of each node is known to the public in PBFT and Ripple, thus they are all appropriate for private blockchain or consortium blockchain. Private blockchains and consortium blockchains are not as decentralized. On the other hand, public blockchain is decentralized. Strong consistency and high efficiency of consensus have made them more suitable for some commercial and medical scenarios.

The steady operations of the blockchain systems are due to the consensus protocol. Via a consensus protocol, the nodes settle on a specific value or transaction. Some of the most widely used consensus protocols were discussed. Not just that but we talked about their strengths, weaknesses, and application scenarios through analysis and comparison. From the discussion above, we can conclude that it is important to design a strong consensus protocol that addresses not only good fault tolerance but also how to make it more useful in a suitable application scenario.

CONCLUSION

Blockchain technology is the new hot thing and it's here to stay. Nowadays everyone is deeply interested in blockchain technology. The majority of the investors have started to invest in this relatively new technology because they were intrigued. Computers running the same blockchain software and wanting to share data join together on a network. The data is assembled together into "blocks" for authentication as it is coming into the network. For instance, through people spending and receiving money. The computers vote on the current block of data systematically, typically every few minutes or even every few seconds deciding whether it all looks good or not. The computer votes on the overruled current block again when the next block is submitted. When the computer agrees about the validity of the data it holds, the current block is accepted, and is then added to the system's complete past history of authenticated data blocks. Hence, the data is "chained." A long chain of information is formed as a result. This process is known as the blockchain. Blockchain is also referred to as Distributed Ledger Technology (DLT). The digital ledger technology allows the history of any digital asset unable to change. It is kept transparent by using decentralization and cryptographic hashing.

Blockchains have a lot of applications in various fields. The music industry also relies a lot on this new technology. Blockchain technology is also one of the most used technologies in the game development field. There are a number of games that are based on the blockchain. These games provide the users with an opportunity to gain control over the game. Previously this was not possible as the conventional gaming industry was centralized and the whole control over the game dynamics was in the hand of the developers. Nowadays, there are multiple ways through which a player can earn from these games. Players can purchase the in-game assets as NFTs and then sell across the blockchain marketplaces. CryptoKitties, Decentraland, and Gods Unchained are a few of the most widely used blockchain-based games.

NFTs are tokens that represent your ownership over a digital asset. NFTs are becoming more and more famous with every passing day and people have started to pay attention to it as well, but they are also subjected to a lot of inspection over their carbon footprint. But NFTs are not the only ones to be blamed for this problem. However, Ethereum may use a strenuous way to secure the funds and assets but its improvement is in the works. Upon improvement, the carbon footprint of Ethereum will be 99.95% better. This would make it more energy-efficient than many existing industries.

Another important thing that you have to bear in mind is that the application of blockchain technology could result in a lot of challenges that need to be addressed. Designing the appropriate consensus protocol is a problem. The consensus of blockchain is that all nodes maintain the same distributed ledger. Centralized servers in the past made the consensus protocol easy to implement. That is because all the nodes would be connected to one central server. This could however prove to be a huge problem in distributed systems because every node is both a host and a server. Both these nodes are supposed to exchange information with the other nodes to reach a consensus. The process of consensus could be disrupted due to offline and malicious nodes. Hence, an excellent consensus protocol can tolerate the occurrence of these phenomena and minimize the damage done so the result of the consensus remains unaffected. Another thing one needs to take care of in the consensus protocol is that you need to choose the consensus protocol that is suitable for the type of blockchain being used.

Every type of blockchain has different application scenarios.

You have to do your research well regarding all these protocols and other crucial things in order to choose a better consensus protocol. Blockchain technology at this point in time is blooming and there is a high chance that it will most likely be the next big thing in the future.

Made in United States
North Haven, CT
05 December 2021

12001075R10104